Lecture Notes in Mathematics Vol. 934
ISBN 978-3-540-11562-5

Makoto Sakai

Quadrature Domains

Errata

I. In the definition of piecewise smooth arcs on page 8, we have merely assumed that it can be expressed as the union of a finite number of smooth arcs each of which is the image of the closed interval $[0, 1]$ under a function of class C^1.

However, to prove Lemma 2.4, which discusses the growth rate of the normal derivative of the Green function at a corner, we need the continuity of the derivative on each of smooth arcs as stated on page 14. Hence our smooth arcs should satisfy a stronger condition so that the normal derivative of the Green function is continuous on each of them.

It is enough to replace "under a function of class C^1" in the definition of a smooth arc with "under a function of class C^2". A weaker condition is "under a Lyapunov-Dini smooth function". A function γ of class C^1 on $[0, 1]$ is called a Lyapunov-Dini smooth function if the modulus of continuity

$$\omega(t) = \sup\{|\gamma'(t_2) - \gamma'(t_1)| : t_j \in [0, 1], \ |t_2 - t_1| \leq t\}$$

of the derivative γ' of γ satisfies

$$\int_0^1 \omega(t)\frac{dt}{t} < +\infty.$$

II. We have mainly discussed quadrature domains of positive measures as stated in Introduction. However, we have treated quadrature domains of real measures and quadrature domains of complex measures without their explicit definitions. For examples, Proposition 8.1 and its corollary are for real measures and Proposition 9.4 is for complex measures. The definitions are the same as in Introduction. The definition of a quadrature domain of a complex measure ν for class AL^1 is the following: A nonempty domain Ω is called a quadrature domain of ν for class AL^1 if

(1) ν is concentrated in Ω, namely, $\nu|\Omega^c = 0$;
(2) $\int_\Omega |f| d|\nu| < +\infty$ and $\int_\Omega f d\nu = \int_\Omega f dm$ for every $f \in AL^1(\Omega)$.

III. When we use the argument as in the proof of Theorem 3.5, for example, in the proofs of Lemma 3.6, Theorem 3.7 and Proposition 3.10, we treat not only a finite positive measure ν stated as in Theorem 3.4, but also a measure of form $\nu + \xi$ for some finite positive measure ξ. In the proof of Theorem 3.5, we depart from $\overline{W} \subset \Omega$ and arrive $\overline{W^{(n)}} \subset \Omega$. We modify the measure ν to $\nu^{(n)}$ in this process. We apply our argument to a measure $\nu + \xi$ for a finite positive measure ξ on Ω such that every $s \in SL^1(\Omega)$ has an integral on Ω and modify $\nu + \xi$ to $\nu^{(n)} + \xi$.

page	line	
4	12	$m(\nu) = \|\nu\| \longrightarrow m(\Omega) = \|\nu\|$
5	2, 7	The double integral $\iint (1/\|\zeta - z\|) d\|\nu\|(\zeta) dm(z)$ is taken over $(\operatorname{supp} \nu)^2$.
7	13	$f \in AL^1(R_\alpha) \longrightarrow f \in A(R_\alpha)$
7	↑8	$(\log r) r > 1$ if $r > 1 \longrightarrow (\log r) r > 0$ if $r > 1$
11	↑4	$d(G_j, \partial O_{j-1} \cup \partial O_{j+1}) \longrightarrow d(G_j, \partial O_{j-2} \cup \partial O_{j+1})$
17	2	Put a period at the end of line.
18	4	are $\longrightarrow$ arc
21	6	with repsect to $\longrightarrow$ with respect to
23	↑3	$q \in \partial W' \longrightarrow q' \in \partial W'$
29	12	$W_n \subset W_{n+1}, n=1,2,\cdots, \bigcup W_n = W$ and $\int_{W_1} \nu dm > m(W_1)$.
	$\longrightarrow$	$W_n \subset W_{n+1} \subset \tilde{W}_n \cup W, n = 1, 2, \cdots, \bigcup W_n \supset W$, $m(\bigcup W_n \setminus W) = 0$ and $\int_{W_1} \nu dm > m(W_1)$,
29	13	Let $\tilde{W}_n$ be $\longrightarrow$ where $\tilde{W}_n$ denotes
31	7	that $\nu_1(z) + \nu_2(z) \geq 1$ a.e. $\longrightarrow$ that $\nu_2(z) \geq 0$ a.e. on $\mathbb{C}$, $\nu_1(z) + \nu_2(z) \geq 1$ a.e.
35	13	these lemma $\longrightarrow$ these lemmas
37	↑9	$E_1 = \overline{R_0} \cap W \longrightarrow E_1 = \overline{R_0} \cap \partial W$
37	↑2–1	$\epsilon = \min\{d(\overline{R_0}, \partial R)/10\sqrt{2}, \inf_{q \in \overline{R_0} \cap \partial W} \lambda_S(q)\}. \longrightarrow \epsilon = \inf_{q \in \overline{R_0} \cap \partial W} \lambda_S(q)$.
38	↑3–2	$\epsilon = d(\overline{R_0}, \partial R)/10\sqrt{2}, \longrightarrow \epsilon = \min\{d(\overline{R_0}, \partial R)/10\sqrt{2}, \inf_{q \in \overline{R_0} \cap \partial W} \lambda_S(q)\}$,
39	3–4	$\epsilon = d(\overline{R_1}, \partial R)/10\sqrt{2} \longrightarrow \epsilon = \min\{d(\overline{R_1}, \partial R)/10\sqrt{2}, \inf_{q \in \overline{R_1} \cap \partial W'} \lambda'_S(q)\}$
42	↑9	$\beta_\Omega^{(n)}(\overline{\Delta(r; p)}) = 0 \longrightarrow \beta_\Omega^{(n)}(\overline{\Delta(R; p)}) = 0$
42	↑8	for some $r > 0 \longrightarrow$ for some $R > r > 0$.
48	↑2	if $v(r) = 0. \longrightarrow$ if $v(r) = 0$, and A and B are nonnegative constants.
65	1	The integral $\int g(\zeta; z, \Omega) dm(\zeta)$ is taken over Ω.
68	↑9	angle $V_1 < 2\Pi \longrightarrow$ angle $V_1 < 3\pi/2$
72	13	$\operatorname{disc}\{\Omega(t)\} = [\bigcup_{t \geq 0} \Omega(t) \setminus \Omega(0)] \setminus \bigcup_{t \geq 0} \partial\Omega(t)$
	$\longrightarrow$	$\operatorname{disc}\{\Omega(t)\} = \left(\bigcup_{t \geq 0} \Omega(t) \setminus \Omega(0)\right) \setminus \bigcup_{t \geq 0} \partial\Omega(t)$
79	1	$\operatorname{stag}\{[W(t)]\} \longrightarrow \operatorname{stag}\{[\tilde{W}(t)]\}$
79	↑10	$Q(\{\chi_\Omega m + \nu(t) - \nu(0)\}, F) \longrightarrow Q(\{\chi_{\Omega(0)} m + \nu(t) - \nu(0)\}, F)$
80	↑9	$Q(0) \in Q(\nu(0), HL^1) \longrightarrow \Omega(0) \in Q(\nu(0), HL^1)$

84	↑1	contradict ⟶ contradicts
95	10	Lemma 6.8, ⟶ Lemma 7.1,
105	8	$Q(\nu, AL') \longrightarrow Q(\nu, AL^1)$
105	↑7–6	$\lvert a_1\rvert \leq \sqrt{2/\pi}(a_0/3)^{3/2} \longrightarrow a_1 = \pi b_1^2 \overline{b_2}$
106	↑10	$\overline{u_j}(x_1, x_2) = \longrightarrow \overline{u_j}(x_1, x_2, t) =$
106	↑9	$(1/d)\int_0^d u_j(x_1, x_2, x_3)dx_3 \longrightarrow (1/d)\int_0^d u_j(x_1, x_2, x_3, t)dx_3$
108	3	For every $z \in C_1(t)$ with angle $V_1 < \pi$, ⟶ For every $z \in C_j(t)$,
108	5	$U \cap \partial\Omega(s)$ is connected ⟶ $U \cap \partial\Omega(s)$ consists of j connected components
109	↑8	every harmonic function on $\overline{\Omega(\tau)}$ ⟶ every harmonic function h on $\overline{\Omega(\tau)}$
110	8	$\mathrm{stat}\{\Omega(t)\} \subset \{z \in C_1(0) \mid \mathrm{angle} V_1 \leq \pi/2\} \subset C_1(0)$ $\longrightarrow \mathrm{stat}\{\Omega(t)\} \subset \{z \in C_1(0) \mid \mathrm{angle} V_1 \leq \pi/2\} \cup \{z \in C_2(0) \mid \mathrm{angle} V_j \leq \pi/2, j = 1, 2\}$
110	↑8	onto ⟶ into
115	5	minimum open set in $Q(\nu, SL^1)$ ⟶ minimum open set Ω in $Q(\nu, SL^1)$
116	↑12	$[G] \longrightarrow G^c$
116	↑4	$\hat{\chi}_{\tilde{G}(t_1)}(\zeta) \longrightarrow \hat{\chi}_{[\tilde{G}(t_1)]}(\zeta)$
117	4	$\varphi'(\zeta_0) > 0 \longrightarrow \varphi'(\zeta) > 0$
122	5	Let G be connected component ⟶ Let G be a connected component
122	12	$\theta(x, G) \longrightarrow \theta(x_i, G)$
124	2, 3	$-(-1)^{n+1}i\} \longrightarrow -(-1)^{n+1}(-1+i)\}$

Lecture Notes in Mathematics

Edited by A. Dold and B. Eckmann

934

Makoto Sakai

Quadrature Domains

Springer-Verlag
Berlin Heidelberg New York 1982

Author
Makoto Sakai
Department of Mathematics, Tokyo Metropolitan University
Fukasawa, Setagaya, Tokyo, 158 Japan

AMS Subject Classifications (1980): 30 E 99, 31 A 99

ISBN 3-540-11562-5 Springer-Verlag Berlin Heidelberg New York
ISBN 0-387-11562-5 Springer-Verlag New York Heidelberg Berlin

Printing and binding: Beltz Offsetdruck, Hemsbach/Bergstr.
2141/3140-543210

CONTENTS

INTRODUCTION

The main purpose of this paper is to show the existence and uniqueness theorems on quadrature domains of positive measures. These can be considered a new type of "the sweeping-out principle" of measures and there are many their applications.

Let ν be a positive Borel measure on the complex plane $\mathbb{C}$. For a domain Ω in $\mathbb{C}$, we denote by $L^1(\Omega)$ the class of all real valued Borel measurable functions on Ω which are integrable with respect to the two-dimensional Lebesgue measure m. Let $F(\Omega)$ be a subclass of $L^1(\Omega)$ such that $f|\Omega \in F(\Omega)$ for every $f \in F(\Omega')$ and every domain Ω' containing Ω.

A nonempty domain Ω is called a quadrature domain of ν for class F if

(1) ν is concentrated in Ω, namely, $\nu(\Omega^c) = 0$.

(2) $\int_\Omega f^+ d\nu < +\infty$ and $\int_\Omega f d\nu \leqq \int_\Omega f dm$

for every $f \in F(\Omega)$, where $f^+ = \max\{f, 0\}$.

If $-f \in F(\Omega)$ for every $f \in F(\Omega)$, then, from (2), we obtain

$$\int_\Omega |f| d\nu < +\infty \quad \text{and} \quad \int_\Omega f d\nu = \int_\Omega f dm$$

for every $f \in F(\Omega)$. Let $AL^1(\Omega)$ be the class of all complex valued analytic integrable functions and set $\mathrm{Re}\, AL^1(\Omega) = \{\mathrm{Re}\, f \mid f \in AL^1(\Omega)\}$. Then Ω is a quadrature domain of ν for $\mathrm{Re}\, AL^1$ if and only if Ω satisfies (1) and, $\int |f| d\nu < +\infty$ and $\int f d\nu = \int_\Omega f dm$ for every $f \in AL^1(\Omega)$. Therefore let us call this domain a quadrature domain for class AL^1. This is nothing but a "classical" quadrature domain of ν.

In this paper, we shall treat quadrature domains for larger classes than Re $AL^1(\Omega)$, namely, we shall treat the class of all harmonic functions in $L^1(\Omega)$ and the class of all subharmonic functions in $L^1(\Omega)$. We denote these classes by $HL^1(\Omega)$ and $SL^1(\Omega)$, respectively. To study quadrature domains for AL^1, it is very important to introduce quadrature domains for these classes.

We mainly concern with quadrature domains with finite area. We denote by $Q(\nu,F)$ the class of all nonempty domains satisfying (1), (2) and

(3) $m(\Omega) < +\infty$.

Quadrature domains with infinite area are discussed in §11. We denote the class by $Q_\infty(\nu,F)$. Since Re $AL^1(\Omega) \subset HL^1(\Omega) \subset SL^1(\Omega)$, we have $Q(\nu,SL^1) \subset Q(\nu,HL^1) \subset Q(\nu,AL^1)$ and $Q_\infty(\nu,SL^1) \subset Q_\infty(\nu,HL^1) \subset Q_\infty(\nu,AL^1)$. The existence and uniqueness theorems on quadrature domains are given for class SL^1.

For applications of our theory, we mention here only three of them. By the existence theorem on quadrature domains for class SL^1, we can estimate the Dirichlet integral of the composite function. As a result, we have an estimation of the Gaussian curvature of the span metric (see §12).

From the existence and uniqueness theorem on an increasing family of quadrature domains for class HL^1, we obtain the existence and uniqueness theorem on the weak solution of the Hele-Shaw flows with a free boundary (see §13).

It is known that there is a domain Ω containing the closed interval $[-1,1]$ and satisfying

$$\int_{-1}^{1} f(x)dx = \int_{\Omega} fdm$$

for every $f \in AL^1(\Omega)$ (see §14). Applying the uniqueness theorem on a quadrature domain for class AL^1, we see that the domain is uniquely determined.

I would like to express here my hearty thanks to Professor H. S. Shapiro for a valuable conversation with him and for his suggestion to the problem of the Hele-Shaw flows with a free boundary.

CHAPTER I. CONSTRUCTION OF QUADRATURE DOMAINS

§1. Elementary properties and examples

At first we note elementary properties of quadrature domains. Let $\Omega \in Q(\nu,F)$ (resp. $\Omega \in Q_\infty(\nu,F)$). If $\Omega \subset \Omega'$ and $m(\Omega'\setminus\Omega) = 0$, then $\Omega' \in Q(\nu,F)$ (resp. $\Omega' \in Q_\infty(\nu,F)$). Hence the areal maximal domain

$$[\Omega] = \{z \in \mathbb{C} \mid m(\Delta(r;z)\setminus\Omega) = 0 \quad \text{for some } r > 0\}$$

of Ω is also in $Q(\nu,F)$ (resp. in $Q_\infty(\nu,F)$), where $\Delta(r;z)$ denotes the open disc with radius r and center at z.

If $\Omega \in Q(\nu,F)$ and $\pm 1 \in F(\Omega)$, then $\|\nu\| = \int d\nu = m(\Omega) < +\infty$. Hence it is necessary to assume $\|\nu\| < +\infty$ for the existence of Ω in $Q(\nu,F)$. We frequently use the equation $m(\nu) = \|\nu\|$ for every $\Omega \in Q(\nu,F)$ and for every class F considered in this paper.

Let E be a Borel measurable set with finite area. Then $\Omega \in Q(\chi_E m,F)$ for every domain Ω with $m(E \,\Delta\, \Omega) = 0$. On the other hand, if $-1 \in F(\Omega)$, then $\Omega \in Q(\chi_E m,F)$ implies $m(E \,\Delta\, \Omega) = 0$. Therefore $\Omega \in Q(\chi_E m,F)$ if and only if $m(E \,\Delta\, \Omega) = 0$ for our classes F.

Next we give here some properties of the Cauchy transforms of measures without proof. Let ν be a complex measure. The Cauchy transform $\hat{\nu}$ of ν is the function defined by

$$\hat{\nu}(z) = \int \frac{d\nu(\zeta)}{\zeta - z} .$$

The Cauchy transform $\hat{\nu}$ is absolutely convergent for almost all z if $\iint (1/|\zeta - z|) d|\nu|(\zeta) dm(z) < +\infty$ and is analytic on the complement of supp ν. For $\nu \in L^1(\mathbb{C}) \cap L^\infty(\mathbb{C})$, we abbreviate $\widehat{\nu m}$ to $\hat{\nu}$. We summarize here properties of the Cauchy transforms. For further properties we can find in Garnett [10, Chapter II].

(1) If ν is a complex measure with $\iint (1/|\zeta - z|) d|\nu|(\zeta) dm(z) < +\infty$, then

$$\frac{\partial \hat{\nu}}{\partial \bar{z}} = -\pi\nu$$

in the sense of distributions.

(2) Let ν be a complex measure with compact support and let φ be the one-to-one conformal mapping defined in a neighborhood of its support. Then

$$\hat{\nu} \circ \varphi^{-1} - \left(\frac{1}{\varphi^{-1\prime}}(\nu \circ \varphi^{-1})\right)^{\wedge}$$

can be extended analytically onto the image of the neighborhood under φ.

(3) If $\nu \in L^1(\mathbb{C}) \cap L^\infty(\mathbb{C})$ and ν is real nonnegative almost everywhere on $\mathbb{C}$, then

$$\|\hat{\nu}\|_\infty \leqq \sqrt{\pi} \sqrt{\|\nu\|_1 \|\nu\|_\infty}.$$

If ν is real, then the inequality holds replacing $\sqrt{\pi}$ by $\sqrt{2\pi}$. For an arbitrary $\nu \in L^1(\mathbb{C}) \cap L^\infty(\mathbb{C})$, we obtain the inequality replacing $\sqrt{\pi}$ by $2\sqrt{\pi}$.

Now we give simple but important examples.

Example 1.1. Let δ_0 be the Dirac measure at the origin 0 and let $\Omega \in Q(\pi\delta_0, AL^1)$. Since $m(\Omega) < +\infty$, $\hat{\chi}_\Omega$ is continuous on $\mathbb{C}$ and $1/(z-\zeta) \in AL^1(\Omega)$ for every $\zeta \in \Omega^c$. Hence $\hat{\chi}_\Omega(\zeta) = -\pi/\zeta$ on the boundary $\partial\Omega$ of Ω. Since $\partial\hat{\chi}_\Omega/\partial\bar{z} = -\pi$ in Ω, we can find a function f analytic on Ω and continuous on the closure $\bar{\Omega}$ of Ω such that $\hat{\chi}_\Omega(z) = -\pi\bar{z} + f(z)$ on $\bar{\Omega}$. Hence $\zeta f(\zeta) = \pi(|\zeta|^2-1)$ on $\partial\Omega$. We shall show in Theorem 6.4 that Ω is bounded, and so $\zeta f(\zeta)$ is bounded. Since its boundary values are real, it must be a constant. Since Ω contains 0 and f is analytic at 0, the constant is equal to zero. Hence $|\zeta| = 1$ on $\partial\Omega$ and so $Q(\pi\delta_0, AL^1) = \{\Delta(1;0)\}$. The mean-value property of harmonic functions and the sub-mean-value property of subharmonic functions imply $Q(\pi\delta_0, SL^1) = Q(\pi\delta_0, HL^1) = \{\Delta(1;0)\}$.

This is the simplest example of quadrature domains. Because of its importance, we give here another proof. As we have seen above, $\hat{\chi}_\Omega(\zeta) = -\pi/\zeta$ on Ω^c. By the inequality of the Cauchy transform given in (3), we have $\pi/|\zeta| \leq \sqrt{\pi m(\Omega)} = \pi$ on Ω^c. Hence $\Delta(1;0) \subset \Omega$. This implies $\Omega = \Delta(1;0)$, because $m(\Omega) = \pi = m(\Delta(1;0))$ and Ω and $\Delta(1;0)$ are both open. This short proof is essentially the same one given by Kuran [13].

Example 1.2. Let ρ be the measure on $\{e^{i\theta} \mid \theta \in \mathbb{R}\}$ defined by $d\rho = (1/(2\pi))d\theta$ and let $\Omega \in Q(t\rho, AL^1)$ for some $t > 0$. Then $\{e^{i\theta} \mid \theta \in \mathbb{R}\} \subset \Omega$, $\widehat{t\rho}(\zeta) = 0$ on $\Delta(1;0)$ and $\widehat{t\rho}(\zeta) = -t/\zeta$ on $\Delta(1;0)^e$. Hence

$$\zeta f(\zeta) = \begin{cases} \pi|\zeta|^2 & \text{on} \quad \Delta(1;0) \cap \partial\Omega \\ \pi|\zeta|^2 - t & \text{on} \quad \Delta(1;0)^e \cap \partial\Omega \end{cases}$$

for some function f analytic on Ω and continuous on $\bar{\Omega}$. Since Ω is bounded (see Theorem 6.4), $f(\zeta) = \alpha/\zeta$ for some constant $\alpha \geqq 0$.

If $\Delta(1;0) \cap \partial\Omega \neq \phi$, then $\Delta(1;0) \cap \partial\Omega = \{\zeta|\ |\zeta| = \sqrt{\alpha/\pi}\}$ and $0 \leqq \alpha < \pi$, because f is analytic on Ω. Hence $\Omega = \{\zeta|\ \sqrt{\alpha/\pi} < |\zeta| < \sqrt{(\alpha+t)/\pi}\} \equiv R_\alpha$, where $0 \leqq \alpha < \pi$ and $\alpha + t > \pi$. If $\Delta(1;0) \cap \partial\Omega = \phi$, then $\alpha = 0$ and $\Omega = \Delta(\sqrt{t/\pi};0)$. It is easy to show that these are quadrature domains. Therefore $Q(t\rho,AL^1) = \{R_\alpha|\ \pi-t < \alpha < \pi\}$ if $0 < t \leqq \pi$ and $Q(t\rho,AL^1) = \{R_\alpha|\ 0 \leqq \alpha < \pi\} \cup \{\Delta(\sqrt{t/\pi};0)\}$ if $t > \pi$.

If $h \in HL^1(R_\alpha)$, then one can find $\gamma \in \mathbb{R}$ and $f \in AL^1(R_\alpha)$ so that $h(z) = \gamma \log|z| + \mathrm{Re}\, f(z)$. Hence $R_\alpha \in Q(t\rho,HL^1)$ if and only if $R_\alpha \in Q(t\rho,AL^1)$ and

$$\int_{\sqrt{\alpha/\pi}}^{\sqrt{(\alpha+t)/\pi}} (\log r) r dr = 0.$$

Since $(\log r)r \leqq 0$ if $0 \leqq r \leqq 1$, $(\log r)r > 1$ if $r > 1$ and $\int_0^1 (\log r) r dr = -1/4$, there is a unique solution $\alpha = \alpha(t)$ of the above equation for t with $0 < t \leqq e\pi$. Therefore $Q(t\rho,HL^1) = \{R_{\alpha(t)}\}$ if $0 < t \leqq \pi$, $Q(t\rho,HL^1) = \{R_{\alpha(t)},\Delta(\sqrt{t/\pi};0)\}$ if $\pi < t \leqq e\pi$ and $Q(t\rho,HL^1) = \{\Delta(\sqrt{t/\pi};0)\}$ if $t > e\pi$.

We shall see in §3 that $Q(t\rho,SL^1) \neq \phi$ and that if $\Omega_j \in Q(t_j\rho,SL^1)$, $j = 1, 2$, and if $t_1 < t_2$, then $\Omega_1 \subset \Omega_2$. Hence, from the above result, we obtain $Q(t\rho,SL^1) = \{R_{\alpha(t)}\}$ if $0 < t < e\pi$,

$Q(e\pi\rho,SL^1) = \{\{z|\ 0 < |z| < \sqrt{e}\},\ \Delta(\sqrt{e};0)\}$ and $Q(t\rho,SL^1) = \{\Delta(\sqrt{t/\pi};0)\}$ if $t > e\pi$.

§2. Domains with quasi-smooth boundaries

In this section we shall define domains with piecewise smooth boundaries and domains with quasi-smooth boundaries. For a finite positive measure μ on $\mathbb{C}$ and a number $N > 0$, we shall define a function $\lambda(z;\mu,N)$ and show $\inf_{z\in\partial W} \lambda(z;\beta,N) > 0$ for a modified measure β of μ if μ is a measure on the closure of a domain W with quasi-smooth boundary such that supp $\mu \not\subset \partial W$. This will be stated in Proposition 2.5. On first reading one could omit this section except for the definition of domains with quasi-smooth boundaries, the definition of a function $\lambda(z;\mu,N)$, the definition of a modified measure β of μ and the statement of Proposition 2.5.

An arc γ is called smooth if it is the image of the closed interval [0,1] under a function of class C^1, we denote it also by γ, such that $d\gamma/dt \neq 0$ on [0,1]. A smooth arc γ is called simple if $\gamma(t_1) \neq \gamma(t_2)$ for every pair of t_1 and t_2 ($\neq t_1$) in [0,1]. We call $\gamma(0)$ and $\gamma(1)$ the ends of γ.

We call a set Γ piecewise smooth arcs if Γ can be expressed as the union of a finite number of smooth simple arcs γ_i, $i = 1,\cdots, m$, such that, for every pair of different γ_i and γ_j, $\gamma_i \cap \gamma_j$ is empty or consists of one point which is the end of both γ_i and γ_j.

A domain W is called with piecewise smooth boundary if ∂W is piecewise smooth arcs.

Let W be a domain with piecewise smooth boundary and let $p \in \partial W$. Let $V_j(r)$, $j = 1,\cdots,n(r)$, be the connected components of $W \cap \Delta(r;p)$. Take $r_p > 0$ so small that, for every r with $0 < r < r_p$ and every j, $j = 1,\cdots,n(p) = n(r_p)$, there is one and only one connected component of $W \cap \Delta(r;p)$ which is contained in $V_j(r_p)$. We denote the component by $V_j(r)$ and define angle V_j by

$$\text{angle } V_j = \lim_{r\to 0} \frac{\ell\left(\overline{V_j(r)} \cap \partial\Delta(r;p)\right)}{r} ,$$

where ℓ denotes the length.

We note here that n(p) and angle V_j, $j = 1,\cdots,n(p)$ are determined independently of the choice of the expression of ∂W. If p is not an end of some expression, then $n(p) \leqq 2$ and angle $V_j = \pi$ for every j with $1 \leqq j \leqq n(p)$. Hence there is only a finite number of points p on ∂W satisfying $n(p) > 2$ or $n(p) \leqq 2$ and angle $V_j \neq \pi$ for some j with $1 \leqq j \leqq n(p)$.

Let γ be a connected component of ∂W. We call γ a smooth simple curve (resp. a piecewise smooth quasi-simple curve) if $n(p) = 1$ and angle $V_1 = \pi$ (resp. $n(p) \leqq 2$) for every $p \in \gamma$. We call γ an analytic simple curve (resp. an analytic quasi-simple curve) if γ is a smooth simple curve (resp. a piecewise smooth quasi-simple curve) and if $\gamma \cap \partial(V_j(r))$ is analytic for sufficiently small $r > 0$, for every $p \in \gamma$ and for every j with $1 \leqq j \leqq n(p)$ except a finite number of points q satisfying $n(q) = 1$ and angle $V_1 = 2\pi$.

We call W a domain with quasi-smooth boundary if W is a domain with piecewise smooth boundary and $\max_{1 \leqq j \leqq n(p)}$ angle $V_j > \pi/2$ for every $p \in \partial W$.

Lemma 2.1. Let W be a domain with piecewise smooth (resp. quasi-smooth) boundary and let c be a fixed point on $\overline{W}$. Then $W \cup \Delta(r;c)$ is a domain with piecewise smooth (resp. quasi-smooth) boundary for almost all $r > 0$.

Proof. Let $\partial W = \cup_{i=1}^{m} \gamma_i$ be an expression of ∂W and set $f_i(t) = |\gamma_i(t)-c|$ and $E_i = \{f_i(t) |\ f_i(t) > 0,\ f_i'(t) = 0\}$. If γ_i and $\partial\Delta(r;c)$ intersect each other infinite times, then $r \in E_i$. Hence, if $r \notin \cup_{i=1}^{m} E_i$, then the boundary of $W \cup \Delta(r;c)$ consists of piecewise smooth arcs and a finite number of points $p_j \in \partial W$ which are ends of some γ_i. Therefore the lemma will be proved if we show $m_1(E_i) = 0$ for each i, where m_1 denotes the one-dimensional Lebesgue measure. This is evident from the following inequality:

$$m_1(E_i) \leqq \int_{\{t|f_i(t)>0, f_i'(t)=0\}} |f_i'(t)| dt = 0.$$

Let μ be a finite positive measure on $\mathbb{C}$. For a fixed number $N > 0$, we define a function $\lambda(z) = \lambda(z;\mu,N)$ on $\mathbb{C}$ by

$$\lambda(z) = \sup\{r \geqq 0 |\ \mu(\overline{\Delta(r;z)}) \geqq N\pi r^2\},$$

where $\overline{\Delta(r;z)} = \{z\}$ if $r = 0$. Then, it is easy to show that

(1) λ is nonnegative upper semicontinuous on $\mathbb{C}$.

(2) $\mu(\overline{\Delta(\lambda(z);z)}) = N\pi(\lambda(z))^2$.

Next we show

Lemma 2.2. Let μ be a finite positive measure on $\mathbb{C}$ and set $\lambda(z) = \lambda(z;\mu,N)$. Then the following are equivalent:

(1) $\lambda = 0$ on $\operatorname{supp} \mu$.

(2) $\lambda = 0$ on $\mathbb{C}$.

(3) μ is absolutely continuous with respect to m and its Radon-Nikodym derivative $d\mu/dm$ satisfies

(a) $\|d\mu/dm\|_\infty \leqq N$.

(b) There are no open discs Δ such that $d\mu/dm = N$ a.e. on Δ.

Proof. It is evident that (3) implies (2) and (2) implies (1). Assume $\lambda = 0$ on $\operatorname{supp} \mu$ and let $d\mu = d\mu_s + (d\mu/dm)dm$, where μ_s is a singular measure with respect to m.

Let A be a Borel subset of $\operatorname{supp} \mu$ such that $m(A) = 0$ and $\mu_s(E) = \mu(A \cap E)$ for every Borel set E on $\mathbb{C}$. Let O be an open set containing A and let $\{O_j\}_{j=1}^\infty$ be a sequence of relatively compact open subsets of O such that $O = \cup_{j=1}^\infty O_j$ and $\overline{O}_j \subset O_{j+1}$. Set $G_1 = O_1$ and $G_j = O_j \backslash O_{j-1}$, $j = 2, 3, \cdots$, and set $d_j = d(G_j, \partial O_{j+1})$, $j = 1, 2$, and $d_j = d(G_j, \partial O_{j-1} \cup \partial O_{j+1})$, $j = 3, 4, \cdots$. Cover G_j by closed squares S_{jk}, $k = 1, \cdots, k_j$ with sides of length $d_j/2$ such that $G_j \cap S_{jk} \neq \phi$ and $S_{jk}^\circ \cap S_{j\ell}^\circ = \phi$ if $k \neq \ell$. Set $S_j = \cup_{k=1}^{k_j} S_{jk}$.

If $p \in A \cap S_{jk}$, then $\mu(A \cap S_{jk}) \leqq \mu(S_{jk}) \leqq \mu(\overline{\Delta(d_j/\sqrt{2};p)}) < N\pi(d_j/\sqrt{2})^2 = 2N\pi m(S_{jk})$. Hence $\mu(A \cap G_j) \leqq \mu(A \cap S_j) < 2N\pi m(S_j)$. Therefore $\mu_s(\mathbb{C}) = \mu(A \cap O) = \mu(A \cap \cup G_j) = \Sigma\mu(A \cap G_j) < 6N\pi m(O)$. Since $m(A) = 0$, this implies $\mu_s(\mathbb{C}) = 0$.

Next we show (a) and (b). If $p \in \text{supp}\,\mu$, then $\mu(\Delta(r;p)) \leqq \mu(\overline{\Delta(r;p)}) < N\pi r^2$ and so $d\mu/dm \leqq N$ a.e. on supp μ. Since $d\mu/dm = 0$ a.e. on $(\text{supp}\,\mu)^c$, (a) holds. If there is an open disc $\Delta(r;c)$ such that $d\mu/dm = N$ a.e. on $\Delta(r;c)$, then $c \in \text{supp}\,\mu$ and $\lambda(c) \geqq r$. This is a contradiction. Hence (3) holds.

<u>Lemma 2.3</u>. Let f be a continuous mapping from piecewise smooth arcs $\Gamma = \cup_{i=1}^{m}\gamma_i$ into $[0,+\infty]$ such that $\int_\Gamma f ds < +\infty$ and $Z = \{p \in \Gamma \mid f(p) = 0\}$ is a finite set, where ds denotes the line element of Γ. Let μ be the measure defined by $\mu(E) = \int_{E\cap\Gamma} f ds$ for every Borel set E in $\mathbb{C}$. If, for every $p \in Z$, there is an arc γ_i containing p and a connected component γ_p of $\gamma_i\setminus\{p\}$ such that

$$\liminf_{\gamma_p \ni q \to p} \frac{f(q)}{|q-p|} > 8N\pi,$$

then $\inf_{z\in\Gamma} \lambda(z;\mu,N) > 0$.

<u>Proof</u>. Let $Z = \{p_1,\cdots,p_n\}$. For every p_j, there is a number $r_j > 0$ such that $f(q) \geqq 8N\pi|q-p_j|$ on $\gamma_{p_j} \cap \overline{\Delta(r_j;p_j)}$ and $\gamma_{p_j} \cap \partial\Delta(r_j;p_j) \neq \phi$. If $p \in \Gamma \cap \overline{\Delta(r_j;p_j)}$, then

$$\mu\left(\overline{\Delta(2r_j;p)}\right) \geqq \int_{\gamma_{p_j}\cap\overline{\Delta(r_j;p_j)}} f ds$$

$$\geqq 8N\pi\int_0^{r_j} rdr$$

$$= 4N\pi r_j^2.$$

Hence $\lambda(p;\mu,N) \geqq 2r_j$ on $\Gamma \cap \overline{\Delta(r_j;p_j)}$.

By using an argument similar to Lemma 2.1, we can find ρ_j so that $0 < \rho_j \leqq r_j$ and $\Gamma\setminus\cup_{j=1}^n \Delta(\rho_j;p_j)$ is piecewise smooth arcs. Since f is continuous, there is $\alpha > 0$ and f satisfies $f \geqq \alpha$ on $\Gamma\setminus\cup_{j=1}^n \Delta(\rho_j;p_j)$. Let ℓ be the minimum length of connected components of $\Gamma\setminus\cup_{j=1}^n \Delta(\rho_j;p_j)$ and set $\lambda_0 = \min\{\alpha/N\pi, \ell\}$. Then $\lambda_0 > 0$ and

$$\mu\left(\overline{\Delta(\lambda_0;p)}\right) = \int_{\overline{\Delta(\lambda_0;p)}\cap\Gamma} f ds \geqq \alpha\lambda_0 \geqq N\pi\lambda_0^2$$

for every $p \in \Gamma\setminus\cup_{j=1}^n \Delta(\rho_j;p_j)$. Hence $\lambda(p;\mu,N) \geqq \min\{\lambda_0, 2r_1, \cdots, 2r_n\}$ on Γ. This completes the proof.

The boundary of a domain W with piecewise smooth boundary is, by definition, piecewise smooth arcs, and so regular with respect to the Dirichlet problem. Hence the Green function $g(z;\zeta,W)$ on W with pole at $\zeta \in W$ exists and can be extended continuously onto ∂W. The Green function on W is characterized by the following:

(1) $g(z;\zeta,W)$ is harmonic in $W\setminus\{\zeta\}$.

(2) $g(z;\zeta,W) - \log(1/|z-\zeta|)$ is harmonic in a neighborhood of ζ.

(3) $g(z;\zeta,W) = 0$ on ∂W.

We divide ∂W into three parts Γ_1, Γ_2 and E defined by $\Gamma_1 = \{p \in \partial W \mid n(p) = 1 \text{ and angle } V_1 = \pi\}$, $\Gamma_2 = \{p \in \partial W \mid n(p) = 2 \text{ and angle } V_1 = \text{angle } V_2 = \pi\}$ and $E = \partial W \setminus (\Gamma_1 \cup \Gamma_2)$.

If $p \in \Gamma_1$, then $g(z;\zeta,W)$ can be extended smoothly onto p. More precisely, for a sufficiently small $r > 0$, there is a C^1-function G on $\Delta(r;p)$ such that $G(z) = g(z;\zeta,W)$ on $V_1(r) = W \cap \Delta(r;p)$. We denote by $\partial g(p;\zeta,W)/\partial n$ the outer normal derivative of G at p. If $p \in \Gamma_2$, then, for a sufficiently small $r > 0$, there are C^1-functions G_1 and G_2 on $\Delta(r;p)$ such that $G_1(z) = g(z;\zeta,W)$ on $V_1(r)$ and $G_2(z) = g(z;\zeta,W)$ on $V_2(r)$. In this case we set $\partial g(p;\zeta,W)/\partial n = \partial G_1(p)/\partial n_1 + \partial G_2(p)/\partial n_2$, where $\partial G_j(p)/\partial n_j$ denotes the outer normal derivative of G_j at p with respect to $V_j(r)$. It is easy to show that $-\partial g(p;\zeta,W)/\partial n$ is continuous and positive on $\Gamma_1 \cup \Gamma_2$.

If $p \in E$, then p is an end of some smooth simple arc for any expression of ∂W. Let p be an end of a smooth simple arc γ on ∂W. We define $\theta = \theta(p,\gamma)$ by $\theta = \text{angle } V_j$ if there is only one $V_j(r)$ satisfying $(\gamma\setminus\{p\}) \cap \partial V_j(r) \neq \phi$, and $\theta = \max\{\text{angle } V_{j_1}, \text{angle } V_{j_2}\}$ if there are two $V_{j_1}(r)$ and $V_{j_2}(r)$ satisfying $(\gamma\setminus\{p\}) \cap \partial V_{j_k}(r) \neq \phi$, $k = 1, 2$, for r with $0 < r < r_p$.

By using the notation above, we have

Lemma 2.4. Let W be a domain with piecewise smooth boundary and let p be an end of some arc γ_i of $\partial W = \cup_{i=1}^{m}\gamma_i$. If $\theta = \theta(p,\gamma_i) > 0$, then

$$\liminf_{\gamma_i\setminus\{p\}\ni q\to p} \frac{-\frac{\partial g(q;\zeta,W)}{\partial n}}{|q-p|^{\pi/\theta-1}} > 0.$$

Proof. It is sufficient to show that

$$\lim_{\gamma_i\setminus\{p\}\ni q\to p} \frac{-\frac{\partial g(q;\zeta,W)}{\partial n}}{|q-p|^{\pi/\theta_j-1}}$$

exists and is not equal to zero, where θ_j = angle V_j, $(\gamma_i\setminus\{p\}) \cap \partial V_j(r) \neq \phi$ for r with $0 < r < r_p$ and $\partial g(q;\zeta,W)/\partial n$ denotes the outer normal derivative of g at q with respect to $V_j(r)$.

Let C be the component of ∂W containing γ_i and let W_1 be the domain surrounded by C such that $W_1 \supset W$. Take $r > 0$ so small that $r < r_p$ and there are no ends in $\overline{\Delta(r;p)}\setminus\{p\}$. From W_1 and $\Delta(r;p)$, by identifying points in $V_j(r)$, we can construct a simply connected planar Riemann surface.

Let φ be a one-to-one conformal mapping from the surface onto the unit disc satisfying $\varphi(p) = 0$. Since $\varphi(W_1)$ is simply connected, we can take the single-valued branch $\Psi(w)$ of w^{π/θ_j} on $\varphi(W_1)$. Consider the composition $\Psi\circ\varphi$ from W_1 onto a Riemann surface R. We denote by $(\Psi\circ\varphi)(W)$ the image of W under $\Psi\circ\varphi$. Since

$$g(q;\zeta,W) = g\big((\Psi\circ\varphi)(q);(\Psi\circ\varphi)(\zeta),(\Psi\circ\varphi)(W)\big),$$

we have

$$\frac{\partial g(q;\zeta,W)}{\partial n} = |(\Psi\circ\varphi)'(q)|\frac{\partial g\big((\Psi\circ\varphi)(q);(\Psi\circ\varphi)(\zeta),(\Psi\circ\varphi)(W)\big)}{\partial n}$$

for $q \in \gamma_i\backslash\{p\}$. Since $\Psi'(w) = (\pi/\theta_j)w^{\pi/\theta_j-1}$ and the boundary of $(\Psi\circ\varphi)(W)$ is smooth at $(\Psi\circ\varphi)(p) \in R$,

$$\lim_{\gamma_i\backslash\{p\}\ni q\to p} \frac{-\frac{\partial g(q;\zeta,W)}{\partial n}}{|\varphi(q)|^{\pi/\theta_j-1}}$$

exists and is not equal to zero. Hence the same holds for

$$\lim_{\gamma_i\backslash\{p\}\ni q\to p} \frac{-\frac{\partial g(q;\zeta,W)}{\partial n}}{|q-p|^{\pi/\theta_j-1}}.$$

Let W be a domain and denote by $HL^\infty(W)$ the Banach space of bounded harmonic functions h on W with norm $\|h\|_\infty = \sup_{z\in W} |h(z)|$. If $h \in HL^\infty(W)$ can be extended continuously onto ∂W, then we say that h belongs to HC(W) and also denote by h its continuous extension. If ∂W is compact and if every point in ∂W is regular with respect to the Dirichlet problem, then the mapping $h \mapsto h|\partial W$ is an isometric isomorphism of HC(W) onto $C(\partial W)$, where $C(\partial W)$ denotes the Banach space of continuous functions on ∂W with norm $\|\cdot\|_\infty$. Hence, for every finite measure μ on $\overline{W}$, there is a unique measure $\beta = \beta(\mu,W)$ on ∂W satisfying

$$\int_{\overline{W}} h\,d\mu = \int_{\partial W} h\,d\beta$$

for every $h \in HC(W)$.

If W is a domain with piecewise smooth boundary and if δ_ζ is the Dirac measure at ζ in W, then

$$\int_{\overline{W}} h\,d\delta_\zeta = h(\zeta) = \int_{\partial W} h(z)\cdot -\frac{1}{2\pi}\frac{\partial g(z;\zeta,W)}{\partial n}\,ds.$$

Hence $d\beta(\delta_\zeta,W) = -(1/(2\pi))(\partial g(z;\zeta,W)/\partial n)ds$

Next we define a function $M_\alpha\nu$. For a finite measure ν and a number $\alpha > 0$, we define a function $M_\alpha\nu$ by $(M_\alpha\nu)(z) = \nu(\Delta(\alpha;z))/\pi\alpha^2$. Then $M_\alpha\nu \in L^1(\mathbb{C}) \cap L^\infty(\mathbb{C})$. If W is a domain such that supp $\nu \subset W$ and $d(\text{supp}\,\nu,\partial W) \geqq \alpha$, then $\int_W h\,d\nu = \int_W h(M_\alpha\nu)dm$ for every $h \in HL^1(W)$.

By using the above notation we have

<u>Proposition 2.5</u>. Let W be a domain with quasi-smooth boundary. For a finite positive measure μ on $\overline{W}$, we set $\beta = \beta(\mu,W)$ and $\lambda(z) = \lambda(z;\beta,N)$. If supp μ is not contained in ∂W, then $\inf_{z\in\partial W} \lambda(z) > 0$.

<u>Proof</u>. Take a small disc $\Delta(r;c)$ in W such that $\mu(\Delta(r;c)) > 0$ and $\Delta(3r;c) \subset W$. Let $\nu = \mu|\Delta(r;c)$ and consider the function $M_{2r}\nu$. Then $(M_{2r}\nu)(z) \geqq \mu(\Delta(r;c))/(4\pi r^2) > 0$ on $\Delta(r;c)$ and

$$\int_{\overline{W}} h\,d\mu = \int_{\Delta(r;c)} h\,d\mu + \int_{\overline{W}\setminus\Delta(r;c)} h\,d\mu$$

$$= \int_W h(M_{2r}\nu)dm + \int_{\overline{W}\setminus\Delta(r;c)} h\,d\mu$$

for every $h \in HC(W)$. Hence, to prove the proposition, without loss of generality we may assume that $d\mu = \alpha\chi_{\Delta(r;c)}dm$, where α is a positive constant and $\Delta(r;c) \subset W$. In this case, we have $d\beta(\mu,W) = -(\alpha r^2/2)(\partial g(z;c,W)/\partial n)ds$.

Let e be an end on ∂W and set $\theta = \max_{1 \leqq j \leqq n(e)}$ angle V_j. Since ∂W is quasi-smooth, $\theta > \pi/2$. Hence, by Lemma 2.4,

$$\lim_{\gamma\setminus\{e\}\ni q\to e} \frac{-\frac{\partial g(q;c,W)}{\partial n}}{|q-e|} = +\infty$$

for some smooth simple are $\gamma \ni e$. Therefore by Lemma 2.3, we have $\inf_{z\in\partial W} \lambda(z) > 0$.

§3. Modifications of positive measures

The existence and uniqueness theorem on quadrature domains for class SL^1 will be proved in this section. Lemmas 3.1 and 3.2 are important. They furnish a key for the solution of the problem.

Lemma 3.1. Let μ be a finite positive measure on $\mathbb{C}$ satisfying supp $\mu \subset \overline{\Delta(\lambda;c)}$ and $\mu(\overline{\Delta(\lambda;c)}) \geqq 144\pi\lambda^2$ for a positive number λ. Then, for every number r with $8\lambda \leqq r \leqq 9\lambda$, there is an L^∞-function $f(z) = f(z;\mu,\Delta(r;c))$ on $\mathbb{C}$ such that

(1) $f(z) \geqq 1$ on $\Delta(r;c)$ and $f(z) = 0$ on $\Delta(r;c)^c$.

(2) $\int_{\overline{\Delta(\lambda;c)}} s d\mu \leqq \int_{\Delta(r;c)} sfdm$ for every $s \in SL^1(\Delta(r;c))$.

Proof. Since every subharmonic function is locally bounded from above, $\int_{\overline{\Delta(\lambda;c)}} s d\mu$ has a meaning. Consider the function $M_{4\lambda}\mu$ which was defined before Proposition 2.5. Then $(M_{4\lambda}\mu)(z) \geqq 9$ on $\Delta(3\lambda;c)$, $(M_{4\lambda}\mu)(z) = 0$ on $\Delta(5\lambda;c)^c$ and

$$\int s\,d\mu \leqq \int_{\Delta(5\lambda;c)} s(M_{4\lambda}\mu)\,dm$$

for every $s \in SL^1(\Delta(5\lambda;c))$, because μ is positive. Let x be the solution of the equation

$$(x-1)\pi(3\lambda)^2 = \pi\left(r^2-(3\lambda)^2\right).$$

Since $8\lambda \leqq r \leqq 9\lambda$, x satisfies $8^2/9 \leqq x \leqq 9$. Set

$$f(z) = \begin{cases} (M_{4\lambda}\mu)(z) - x + 1 & \text{on} \quad \Delta(3\lambda;c) \\ (M_{4\lambda}\mu)(z) + 1 & \text{on} \quad \Delta(r;c)\setminus\Delta(3\lambda;c) \\ 0 & \text{on} \quad \Delta(r;c)^c. \end{cases}$$

Then f is the desired function.

Lemma 3.2. Let μ be a finite positive measure with compact support K. For a fixed number $N > 0$, let $\lambda(z) = \lambda(z;\mu,N)$ be the function on $\mathbb{C}$ defined before Lemma 2.2. Suppose $\lambda > 0$ on K and set $U = \cup_{q\in K}\,\Delta(3\lambda(q);q)$. Then

$$\int_K s\,d\mu \leqq 36N\int_U s\,dm$$

for every nonnegative s in $SL^1(U)$.

Proof. To prove the lemma, we define a function $\Lambda(p)$ on $\mathbb{C}$ by

$$\Lambda(p) = \sup_{q\in K}\,\{\lambda(q) - \frac{|q-p|}{3}\}.$$

It is easy to show that

(1) In the above equality, we can replace sup by max, namely, there is a point in K where the supremum is attained.

(2) Λ is continuous on $\mathbb{C}$.

(3) $0 < \lambda \leqq \Lambda$ on K.

(4) $\cup_{q\in K} \Delta(\Lambda(q);q) \subset U$.

(5) $K_p \equiv \{q \in K \mid |p-q| \leqq \Lambda(q)\}$ is closed for every fixed point $p \in \mathbb{C}$. In particular, if $K_p \neq \phi$, there is a point $q_1 \in K_p$ such that $|p-q_1| = \sup_{q\in K_p} |p-q|$.

(6) If $q_1 \in K_p$, then

(a) $\mu(\overline{\Delta(|p-q_1|;p)}) \leqq 4N\pi\Lambda(q_1)^2$.

(b) $\Lambda(q) \geqq \Lambda(q_1)/3$ on $\overline{\Delta(|p-q_1|;p)}$.

Let s be a nonnegative function in $SL^1(U)$. By using (3), (4), the sub-mean-value property of subharmonic functions and the Fubini theorem, we have

$$\int_K s(q)d\mu(q) \leqq \int_K \left\{\frac{1}{\pi\Lambda(q)^2}\int_{\Delta(\Lambda(q);q)} s(p)dm(p)\right\}d\mu(q)$$

$$= \int_U \left\{\int_{\{q\in K \mid |q-p|<\Lambda(q)\}} \frac{d\mu(q)}{\pi\Lambda(q)^2}\right\} s(p)dm(p).$$

For every p with $K_p \neq \phi$, take q_1 stated as in (5). Then $\{q \in K \mid |q-p| < \Lambda(q)\} \subset K_p \subset \overline{\Delta(|p-q_1|;p)}$ and so, by (6),

$$(3.1) \qquad \int_{\{q\in K \mid |q-p|<\Lambda(q)\}} \frac{d\mu(q)}{\pi\Lambda(q)^2} \leqq \int_{\overline{\Delta(|p-q_1|;p)}} \frac{d\mu(q)}{\pi\Lambda(q)^2}$$

$$\leqq \frac{4N\pi\Lambda(q_1)^2}{\pi\left(\frac{1}{3}\Lambda(q_1)\right)^2} = 36N.$$

Since s is nonnegative, we have our lemma.

Now let us study the modification of positive measures.

Proposition 3.3. Let W be a bounded domain with quasi-smooth boundary. Let ν be a finite positive measure on $\overline{W}$ satisfying $d\nu/dm \geqq \chi_W$, where $d\nu/dm$ denotes the Radon-Nikodym derivative of ν with repsect to m. Let $N \geqq 144$ and set $\beta = \beta(\nu-\chi_W m, W)$ and $\lambda(z) = \lambda(z;\beta,N)$.

If $\inf_{z\in\partial W} \lambda(z) > 0$, then one can construct a bounded domain W' with quasi-smooth boundary and a finite positive measure ν' on $\overline{W'}$ satisfying $d\nu'/dm \geqq \chi_{W'}$ and the following conditions:

(1) $\overline{W} \subset W'$.

(2) $\int_{\overline{W}} s\,d\nu \leqq \int_{\overline{W'}} s\,d\nu'$ for every $s \in S(\overline{W'})$,

where $S(\overline{W}')$ denotes the class of all functions subharmonic on an open neighborhood of $\overline{W}'$.

(3) $m(W') \leqq \nu(\overline{W})$.

(4) Set $\beta' = \beta(\nu'-\chi_{W'}m, W')$, $\lambda'(z) = \lambda(z;\beta',N)$, $U = \cup_{q\in\partial W}\Delta(3\lambda(q);q)$ and $U' = \cup_{q\in\partial W'}\Delta(3\lambda'(q);q)$. Then $\inf_{z\in\partial W'}\lambda'(z) > 0$, $U \subset W'$ and $U \cap U' = \phi$.

Proof. Set $E_1 = \partial W$. The upper semicontinuous function λ attains its maximum on the compact set E_1 at a point p_1. Set $E_2 = E_1 \backslash \Delta(2\lambda(p_1);p_1)$. If $E_2 \neq \phi$, then we can again find $p_2 \in E_2$ at which λ attains its maximum on E_2. Define E_3 by $E_3 = E_2 \backslash \Delta(2\lambda(p_2);p_2)$. We can continue this process as long as $E_j \neq \phi$.

Since $\inf_{p\in\partial W} \lambda(p) > 0$, there is a number n such that $E_n \neq \phi$ and $E_{n+1} = \phi$. By the construction, the discs $\Delta(\lambda(p_j);p_j)$, $j = 1,\cdots, n$, are mutually disjoint and $U \subset \cup_{j=1}^{n}\Delta(5\lambda(p_j);p_j)$.

Next we choose r_j with $8\lambda(p_j) \leqq r_j \leqq 9\lambda(p_j)$ so that $W_0 = W \cup \cup_{j=1}^{n}\Delta(r_j;p_j)$ is a domain with quasi-smooth boundary, by using Lemma 2.1.

We define ν_0 by

$$d\nu_0 = \chi_W dm + d\beta|\{\partial W\setminus \overline{\bigcup_{j=1}^{n} \Delta(\lambda(p_j);p_j)}\}$$

$$+ \sum_{j=1}^{n} f\left(z;\beta|\overline{\Delta(\lambda(p_j);p_j)},\ \Delta(r_j;p_j)\right)dm,$$

where f is the function defined in Lemma 3.1. Then $d\nu_0/dm \geqq \chi_{W_0}$ and, by Lemma 3.1,

$$\int_{\overline{W}} s d\nu \leqq \int_{\overline{W}_0} s d\nu_0$$

for every $s \in S(\overline{W}_0)$. Set $\beta_0 = \beta(\nu_0 - \chi_{W_0} m, W_0)$ and $\lambda_0(z) = \lambda(z;\beta_0,N)$.

If $\max_{z\in\partial W_0} \lambda_0(z) \geqq \inf_{z\in\partial W} \lambda(z)$, then choose a point $q_0 \in \partial W_0$ so that $\lambda_0(q_0) \geqq \inf_{z\in\partial W} \lambda(z)$ and choose ρ_0 with $8\lambda_0(q_0) \leqq \rho_0 \leqq 9\lambda_0(q_0)$ so that $W_1 = W_0 \cup \Delta(\rho_0;q_0)$ is a domain with quasi-smooth boundary. Define ν_1, β_1 and λ_1 by

$$d\nu_1 = \chi_{W_0} dm + d\beta_0|\{\partial W_0\setminus\overline{\Delta(\lambda_0(q_0);q_0)}\}$$

$$+ f\left(z;\beta|\overline{\Delta(\lambda(q_0);q_0)},\Delta(\rho_0;q_0)\right)dm,$$

$$\beta_1 = \beta(\nu_1 - \chi_{W_1} m, W_1)$$

and

$$\lambda_1(z) = \lambda(z;\beta_1,N).$$

If $\max_{z\in\partial W_1} \lambda_1(z) \geqq \inf_{z\in\partial W} \lambda(z)$, then, by using the same argument as above, we can choose a point $q_1 \in \partial W_1$ so that $\lambda_1(q_1) \geqq \inf_{z\in\partial W} \lambda(z)$, a number ρ_1 with $8\lambda_1(q_1) \leqq \rho_1 \leqq 9\lambda_1(q_1)$ so that $W_2 = W_1 \cup \Delta(\rho_1;q_1)$ is a domain with quasi-smooth boundary and construct ν_2, β_2 and λ_2.

We continue this process as long as possible. Since $|q_i-q_j| \geqq 8 \inf_{z\in\partial W} \lambda(z)$ if $i \neq j$ and $m(W_j) \leqq \nu(\overline{W})$ for every j, our process must stop after a finite number of times. Therefore there are $k \geqq 0$ satisfying $\max_{z\in\partial W_k} \lambda_k(z) < \inf_{z\in\partial W} \lambda(z)$ and $\max_{z\in\partial W_j} \lambda_j(z) \geqq \inf_{z\in\partial W} \lambda(z)$ for every $j < k$.

Set $W' = W_k$ and $\nu' = \nu_k$. It is easy to show that W' and ν' satisfy (1), (2), (3) and $U \subset W_0 \subset W'$. By Proposition 2.5, it follows that $\inf_{z\in\partial W_j} \lambda_j(z) > 0$ for every $j \leqq k$. Hence $\inf_{z\in\partial W'} \lambda'(z) > 0$.

Finally let us show $U \cap U' = \phi$. Assume $z \in U \cap U'$. Then there are points $q \in \partial W$ and $q' \in \partial W'$ satisfying $|z-q| < 3\lambda(q)$ and $|z-q'| < 3\lambda'(q')$, respectively. Hence $|q-q'| < 3\lambda(q) + 3\lambda'(q') < 6\lambda(q)$. Therefore $q' \in \cup_{j=1}^n \Delta(8\lambda(p_j);p_j)$, because $\cup_{q\in\partial W} \Delta(6\lambda(q);q) \subset \cup_{j=1}^n \Delta(8\lambda(p_j);p_j)$. This implies $q' \in W_0 \subset W'$ and contradicts $q \in \partial W'$. The proof is complete.

<u>Theorem 3.4</u>. Let W be a bounded domain with quasi-smooth boundary. Let ν be a finite positive measure on $\overline{W}$ with $d\nu/dm \geqq$

χ_W and $\int d\nu > m(W)$. Then one can construct a domain $\tilde{W}$ satisfying

(1) $\overline{W} \subset \tilde{W}$.

(2) $\int_{\overline{W}} s d\nu \leqq \int_{\tilde{W}} s dm$ for every $s \in SL^1(\tilde{W})$.

(3) $m(\tilde{W}) = \nu(\overline{W})$.

<u>Proof</u>. Let $N \geqq 144$, $\beta = \beta(\nu - \chi_W m, W)$ and $\lambda(z) = \lambda(z;\beta,N)$. Then, by Lemma 2.2, λ is not identically equal to zero on ∂W. Hence there is a point $p \in \partial W$ such that $\lambda(p) > 0$. Choose r with $8\lambda(p) \leqq r \leqq 9\lambda(p)$ so that $W^{(0)} = W \cup \Delta(r;p)$ is a domain with quasi-smooth boundary. Set

$$d\nu^{(0)} = \chi_W dm + d\beta|\{\partial W \setminus \overline{\Delta(\lambda(p);p)}\} + f\big(z;\beta|\overline{\Delta(\lambda(p);p)}, \Delta(r;p)\big)dm,$$

$$\beta^{(0)} = \beta\big(\nu^{(0)} - \chi_{W^{(0)}} m, W^{(0)}\big)$$

and

$$\lambda^{(0)}(z) = \lambda\big(z;\beta^{(0)},N\big).$$

Then $W \subset W^{(0)}$, $m(W^{(0)}) \leqq \nu(\overline{W})$ and $\int_{\overline{W}} s d\nu \leqq \int_{\overline{W^{(0)}}} s d\nu^{(0)}$ for every $s \in S(\overline{W^{(0)}})$. Moreover, by Proposition 2.5, $\inf_{z \in W^{(0)}} \lambda^{(0)}(z) > 0$.

Next we apply Proposition 3.3 and define $W^{(n)}$ and $\nu^{(n)}$ inductively by $W^{(n)} = (W^{(n-1)})'$ and $\nu^{(n)} = (\nu^{(n-1)})'$, $n = 1, 2, \cdots$. Let $\beta^{(n)} = \beta(\nu^{(n)} - \chi_{W^{(n)}} m, W^{(n)})$, $\lambda^{(n)}(z) = \lambda(z;\beta^{(n)},N)$ and $U^{(n)} = \cup_{q \in \partial W^{(n)}} \Delta(3\lambda^{(n)}(q);q)$ and set $\tilde{W} = \cup_{n=0}^{\infty} W^{(n)}$. Then, by Lemma 3.2,

$$\int_{\overline{W}} s d\nu \leqq \int_{W^{(n)}} s dm + \int_{\partial W^{(n)}} s d\beta^{(n)}$$

$$\leqq \int_{W^{(n)}} s dm + 36N \int_{U^{(n)}} s dm$$

for every nonnegative $s \in SL^1(\tilde{W})$. Since $\{U^{(n)}\}_{n=0}^{\infty}$ is a sequence of mutually disjoint open sets and $\cup_{n=0}^{\infty} U^{(n)} \subset \tilde{W}$, we have

$$\lim_{n\to\infty} \int_{U^{(n)}} s dm = 0,$$

and hence

$$\int_{\overline{W}} s d\nu \leqq \int_{\tilde{W}} s dm$$

for every nonnegative $s \in SL^1(\tilde{W})$.

By Proposition 3.3, (1) and $m(\tilde{W}) \leqq \nu(\overline{W})$ hold. Hence the function 1 is in $SL^1(\tilde{W})$ and so, by the above inequality, we have $\nu(\overline{W}) \leqq m(\tilde{W})$. Therefore (3) holds.

Let $s \in SL^1(\tilde{W})$. For every $x \in \mathbb{R}$, the function $\max\{s,x\} - x$ is nonnegative and belongs to $SL^1(\tilde{W})$. Hence

$$\int_{\overline{W}} (\max\{s,x\}-x) d\nu \leqq \int_{\tilde{W}} (\max\{s,x\}-x) dm$$

for every $x \in \mathbb{R}$. By (3), this implies

$$\int_{\overline{W}} \max\{s,x\} d\nu \leqq \int_{\tilde{W}} \max\{s,x\} dm.$$

Letting $x \downarrow -\infty$, we have the inequality stated in (2).

Next we show the uniqueness theorem.

Theorem 3.5. Let W be a domain and ν be a measure stated as in Theorem 3.4. Take $N \geqq 100e$ and construct a domain $\tilde{W}$ as in Theorem 3.4. Then $\tilde{W}$ is uniquely determined. Moreover, a domain Ω satisfies

(1) $\overline{W} \subset \Omega$.

(2) $\int_{\overline{W}} s d\nu \leqq \int_{\Omega} s dm$ for every $s \in SL^1(\Omega)$.

(3) $m(\Omega) = \nu(\overline{W})$.

if and only if Ω satisfies $\tilde{W} \subset \Omega$ and $m(\Omega \setminus \tilde{W}) = 0$.

Proof. If Ω satisfies $\tilde{W} \subset \Omega$ and $m(\Omega \setminus \tilde{W}) = 0$, then $s|\tilde{W} \in SL^1(\tilde{W})$ for every $s \in SL^1(\Omega)$ and $\int_{\tilde{W}} s dm = \int_{\Omega} s dm$. Hence Ω satisfies from (1) to (3).

Assume next that a domain Ω satisfies (1) and (2). Let us show that Ω satisfies

(1;0) $\overline{W^{(0)}} \subset \Omega$.

(2;0) $\int_{\overline{W^{(0)}}} s d\nu^{(0)} \leqq \int_{\Omega} s dm$ for every $s \in SL^1(\Omega)$.

At first we show $\int_W s dm + \int_{\partial W} s d\beta \leqq \int_{\Omega} s dm$ for every $s \in SL^1(\Omega)$. Let H_s^W be the solution in W of the Dirichlet problem for the boundary function s and set

$$s^*(z) = \begin{cases} H_s^W(z) & \text{on} \quad W \\ s(z) & \text{on} \quad \Omega \setminus W. \end{cases}$$

Then $s^* \in SL^1(\Omega)$ and, by (2),

$$\int_{\overline{W}} s^* d\nu \leqq \int_{\Omega} s^* dm.$$

We add $\int_W sdm - \int_W H_s^W dm$ to both sides and obtain

$$\int_W sdm + \int_{\partial W} sd\beta = \int_W sdm + \int_W H_s^W d(\nu - \chi_W m) + \int_{\partial W} sd\nu \leqq \int_{\Omega} sdm.$$

Next we show (1;0). Take a point $q \notin \Omega$ and consider the subharmonic function $s(z) = \max\{\frac{1}{2} + \log \frac{10\lambda(p)}{|z-q|}, 0\}$, then $s|\Omega \in SL^1(\Omega)$, $s(z) = 0$ on $\Delta(10\sqrt{e}\lambda(p);q)^c$ and

$$\int_{\Omega} sdm \leqq \int_{\Delta(10\sqrt{e}\lambda(p);q)} sdm = 50e\pi\lambda(p)^2.$$

If $|p-q| \leqq r$, then $|z-q| \leqq |z-p| + |p-q| \leqq \lambda(p) + 9\lambda(p) = 10\lambda(p)$ on $\overline{\Delta(\lambda(p);p)}$ and so $s(z) \geqq 1/2$ on $\overline{\Delta(\lambda(p);p)}$. Hence

$$\int_W sdm + \int_{\partial W} sd\beta > \int_{\overline{\Delta(\lambda(p);p)}} sd\beta \geqq \beta(\overline{\Delta(\lambda(p);p)})/2 \geqq 50e\pi\lambda(p)^2.$$

This contradicts the above inequality. Therefore $\overline{\Delta(r;p)} \subset \Omega$ and so (1;0) holds.

Now we show (2;0). Let $H_s^{\Delta(r;p)}$ be the solution in $\Delta(r;p)$ of the Dirichlet problem for the boundary function s and set

$$s^*(z) = \begin{cases} H_s^{\Delta(r;p)}(z) & \text{on} \quad \Delta(r;p) \\ s(z) & \text{on} \quad \Omega \setminus \Delta(r;p). \end{cases}$$

Then $s^* \in SL^1(\Omega)$ and so

$$\int_W s^* dm + \int_{\partial W} s^* d\beta \leqq \int_{\Omega} s^* dm.$$

Set $\Delta = \Delta(r;p)$ and $H = H_s^{\Delta}$. Since $H \geqq s$ and $f(z;\beta|\overline{\Delta(\lambda(p);p)},\Delta) \geqq 1$ on Δ,

$$\int_{\Delta} sfdm - \int_{\Delta} Hfdm \leqq \int_{\Delta} sdm - \int_{\Delta} Hdm.$$

We add these two inequalities and obtain

$$\int_{W} s^*dm + \int_{\partial W\setminus\overline{\Delta(\lambda(p);p)}} s^*d\beta + \int_{\Delta} sfdm \leqq \int_{\Omega} sdm.$$

Hence (2;0) holds.

By using the same argument as above, we see that Ω satisfies

(1;n) $\overline{W^{(n)}} \subset \Omega.$

(2;n) $\int_{\overline{W^{(n)}}} sd\nu^{(n)} \leqq \int_{\Omega} sdm$ for every $s \in SL^1(\Omega)$.

for every n. This implies $\tilde{W} = \cup_{n=0}^{\infty} W^{(n)} \subset \Omega$ and so $m(\Omega\setminus\tilde{W}) = 0$ if Ω satisfies (3).

<u>Lemma 3.6</u>. Let W_j and ν_j, j = 1, 2, be domains and measures stated as in Theorem 3.4, respectively. Let $\tilde{W}_j$ be domains constructed in Theorem 3.4 for W_j and ν_j taking $N \geqq 100e$. If $\nu_1 \leqq \nu_2$, namely, if $\nu_1(E) \leqq \nu_2(E)$ for every Borel set E, then $\tilde{W}_1 \subset \tilde{W}_2$.

<u>Proof</u>. Let $\{W^{(n)}\}$ and $\{\nu^{(n)}\}$ be sequences of domains with quasi-smooth boundaries and measures constructed in the proof of Theorem 3.4 for W_1 and ν_1 taking $N \geqq 100e$, respectively. Here we choose $\{W^{(n)}\}$ so that each $W^{(n)} \cup W_2$ is also a domain with quasi-smooth boundary. Since $\nu_2 - \nu_1 \geqq 0$, we can construct a unique domain $\tilde{W}^{(n)}$ starting from $W^{(n)} \cup W_2$ and $\nu^{(n)} + (\nu_2-\nu_1)$. Since $\nu_2 = \nu_1 + (\nu_2-\nu_1)$, by Theorem 3.5, $\tilde{W}^{(n)} = \tilde{W}_2$. Hence $W^{(n)} \subset \tilde{W}_2$ for every n and so $\tilde{W}_1 \subset \tilde{W}_2$.

In Theorem 3.4, we have proved that $\tilde{W} \in Q(\nu, SL^1)$, because $s^+ = \max\{s,0\} \in SL^1(\tilde{W})$ if $s \in SL^1(\tilde{W})$. Next we study the modification of absolutely continuous measures.

Theorem 3.7. Let ν be an L^∞-function on $\mathbb{C}$ such that $\nu(z) \geqq 1$ a.e. on a domain W with $m(W) < +\infty$, $\nu(z) = 0$ a.e. on W^c and $\int \nu dm > m(W)$. Then there exists a minimum domain $\tilde{W}$ in $Q(\nu m, SL^1)$, namely, $\Omega \in Q(\nu m, SL^1)$ if and only if $\tilde{W} \subset \Omega$ and $m(\Omega \setminus \tilde{W}) = 0$. Moreover, if W is a domain with quasi-smooth boundary, then $\overline{W} \subset \tilde{W}$.

Proof. First we construct a domain $\tilde{W}$. Let $\{W_n\}_{n=1}^\infty$ be a sequence of bounded subdomains of W with quasi-smooth boundaries such that $W_n \subset W_{n+1}$, $n = 1, 2, \cdots$, $\cup W_n = W$ and $\int_{W_1} \nu dm > m(W_1)$. Let $\tilde{W}_n$ be a unique domain according to Theorem 3.5 starting from W_n and $\nu \chi_{W_n} m$. Then, by Lemma 3.6, $\tilde{W}_n \subset \tilde{W}_{n+1}$. Since $m(\tilde{W}_n) \leqq \int \nu dm$ and

$$\int_W s\nu dm \leqq \int_{\tilde{W}_n} s dm + \int_{W \setminus W_n} s\nu dm$$

for every $s \in SL^1(\tilde{W}_n \cup W)$, $\tilde{W} = \cup \tilde{W}_n$ belongs to $Q(\nu m, SL^1)$.

If $\Omega \in Q(\nu m, SL^1)$ and $W \subset \Omega$, then, by using the same argument as in the proof of Theorem 3.5, we have $\tilde{W}_n \subset \Omega$ for every n. Hence $\tilde{W} \subset \Omega$. This implies that $\tilde{W}$ is determined independently of the choice of the sequence $\{W_n\}$.

Let $[W]$ be the areal maximal domain of W. Since $\overline{W}_n \subset \tilde{W}_n$ according to Theorem 3.4, by choosing suitable W_n, we see that $[W] \subset \tilde{W}$. Hence $[W]^\sim \subset \tilde{W}$. Since $W \subset [W] \subset [W]^\sim$, we also have

$\tilde{W} \subset [W]^{\sim}$. Thus $\tilde{W} = [W]^{\sim}$ and $\tilde{W}$ is determined independently of the choice of W, because [W] is determined uniquely by ν.

The proof will be complete if we show that $\tilde{W} \subset \Omega$ for every $\Omega \in Q(\nu m, SL^1)$. If $W \backslash \Omega = \phi$, then we have proved $\tilde{W} \subset \Omega$. Assume $W \backslash \Omega \neq \phi$. Take a subset E of $W \backslash \Omega$ so that $E \cup (W \cap \Omega)$ is a domain. Then $E \cup (W \cap \Omega) \subset E \cup \Omega$ and $E \cup \Omega \in Q(\nu m, SL^1)$. Hence, by the above argument, $\tilde{W} = \{E \cup (W \cap \Omega)\}^{\sim} \subset E \cup \Omega$. Taking E and another E_1 so that $E \cap E_1 = \phi$, we have $\tilde{W} \subset \Omega$.

By Theorem 3.7, we see that the domain $\tilde{W}$ constructed in Theorem 3.4 is the minimum domain in $Q(\nu, SL^1)$.

<u>Proposition 3.8</u>. Let ν_j, $j = 1, 2$, be L^{∞}-functions on $\mathbb{C}$ stated as in Theorem 3.7 and let $\tilde{W}_j$ be the minimum domain in $Q(\nu_j m, SL^1)$. If $\nu_1 \leqq \nu_2$ a.e. on $\mathbb{C}$, then $\tilde{W}_1 \subset \tilde{W}_2$.

<u>Proof</u>. Let W_j be domains such that $\nu_j(z) \geqq 1$ a.e. on W_j and $\nu_j(z) = 0$ a.e. on W_j^c. We may assume $W_1 \subset W_2$. Let $\{W_{1,n}\}$ be a sequence of bounded subdomains of W with quasi-smooth boundaries such that $W_{1,n} \subset W_{1,n+1}$, $n = 1, 2, \cdots$, $\cup W_{1,n} = W_1$ and $\int_{W_{1,1}} \nu_1 dm > m(W_{1,1})$. Then $W_{1,n}^{\sim} \subset \tilde{W}_2$ for every n according to the uniqueness of $\tilde{W}_2$. Hence $\tilde{W}_1 \subset \tilde{W}_2$.

<u>Corollary 3.9</u>. Let ν_j, $j = 0, 1, \cdots$, be L^{∞}-functions on $\mathbb{C}$ stated as in Theorem 3.7 such that $\nu_j \leqq \nu_{j+1}$ a.e. on $\mathbb{C}$ for $j \geqq 1$ and $\nu_j \uparrow \nu_0$ a.e. on $\mathbb{C}$. Let $\tilde{W}_j$ be the minimum domain in $Q(\nu_j m, SL^1)$. Then $\tilde{W}_j \uparrow \tilde{W}_0$.

Proof. By Proposition 3.8, $\tilde{W}_j \uparrow \cup_{j\geqq 1} \tilde{W}_j$ and $\cup_{j\geqq 1} \tilde{W}_j \subset \tilde{W}_0$. Since $\nu_j \uparrow \nu_0$, $\cup_{j\geqq 1} \tilde{W}_j \in Q(\nu_0 m, SL^1)$ and so $\tilde{W}_0 \subset \cup_{j\geqq 1} \tilde{W}_j$. Hence $\tilde{W}_0 = \cup_{j\geqq 1} \tilde{W}_j$.

Proposition 3.10. Let ν_1 be an L^∞-function on $\mathbb{C}$ such that $\nu_1(z) \geqq 1$ a.e. on a domain W_1 with $m(W_1) < +\infty$, $\nu_1(z) = 0$ a.e. on W_1^c and $\int \nu_1 dm > m(W_1)$. Let ν_2 be an L^∞-function on $\mathbb{C}$ such that $\nu_1(z) + \nu_2(z) \geqq 1$ a.e. on a domain W with $m(W) < +\infty$, $\nu_1(z) + \nu_2(z) = 0$ a.e. on W^c. Then $Q(\chi_{\tilde{W}_1} m + \nu_2 m, SL^1) = Q(\nu_1 m + \nu_2 m, SL^1)$.

Proof. Since $\chi_{\tilde{W}_1} + \nu_2 \geqq 1$ a.e. on $\tilde{W}_1 \cup W$ and $\chi_{\tilde{W}_1} + \nu_2 = 0$ a.e. on $(\tilde{W}_1 \cup W)^c$, there is the minimum domain V in $Q(\chi_{\tilde{W}_1} m + \nu_2 m, SL^1)$. If $\Omega \in Q(\chi_{\tilde{W}_1} m + \nu_2 m, SL^1)$, then $\Omega \in Q(\nu_1 m + \nu_2 m, SL^1)$ and so it is sufficient to show that $V \subset \Omega$ for every $\Omega \in Q(\nu_1 m + \nu_2 m, SL^1)$. If $\Omega \in Q(\nu_1 m + \nu_2 m, SL^1)$, then $W_1 \subset \Omega$ and so $\tilde{W}_1 \subset \Omega$. Hence $\tilde{W}_1 \cup W \subset \Omega$ so that $V \subset \Omega$.

Corollary 3.11. Let ν_j, $j = 1, 2$, be L^∞-functions on $\mathbb{C}$ stated as in Theorem 3.7. Then

$$m(\Omega_1 \,\Delta\, \Omega_2) \leqq \int |\nu_1 - \nu_2| dm$$

for every $\Omega_j \in Q(\nu_j m, SL^1)$, $j = 1, 2$.

Proof. We may assume that $\int |\nu_1 - \nu_2| dm > 0$ and Ω_j is the minimum domain $\tilde{W}_j$ in $Q(\nu_j m, SL^1)$. Set $\nu = \min\{\nu_1, \nu_2\}$. If $W_1 \cap W_2$ is a domain, let $\tilde{W}$ be the minimum domain in $Q(\nu \chi_{W_1 \cap W_2} m, SL^1)$.

Then, by Proposition 3.10, $\tilde{W}_j$ is the minimum domain in $Q(\chi_{\tilde{W}}m+\nu_j m-\nu m, SL^1)$. Hence $m(\tilde{W}_j\setminus\tilde{W}) = \int(\nu_j-\nu)dm$ and so $m(\tilde{W}_1 \Delta \tilde{W}_2) \leqq m(\tilde{W}_1\setminus\tilde{W}) + m(\tilde{W}_2\setminus\tilde{W}) = \int|\nu_1-\nu_2|dm$. If $W_1 \cap W_2$ is not connected, we choose an open set O so that $O \cup (W_1 \cap W_2)$ is connected. Then $(\chi_O+\nu_j)(z) \geqq 1$ a.e. on $O \cup W_j$ and $(\chi_O+\nu_j)(z) = 0$ a.e. on $(O \cup W_j)^c$. Letting $m(O) \to 0$, we have

$$\int |\nu_1-\nu_2|dm \geqq m\left((O \cup W_1)^{\sim} \Delta (O \cup W_2)^{\sim}\right) \to m(\tilde{W}_1 \Delta \tilde{W}_2).$$

If $m(W) < +\infty$, then $SL^q(W) \subset SL^p(W)$ for $1 \leqq p \leqq q \leqq +\infty$. Hence $Q(\nu,SL^p) \subset Q(\nu,SL^q)$. Let ν be an L^p function on $\mathbb{C}$ such that $\nu(z) \geqq 1$ a.e. on a domain W with $m(W) < +\infty$, $\nu(z) = 0$ a.e. on W^c and $\int\nu dm > m(W)$. By using the same argument as in Theorems 3.4 and 3.5, we see that if $p > 1$, then there exists the minimum domain in $Q(\nu m,SL^{p'})$, where $\frac{1}{p}+\frac{1}{p'} = 1$, and $Q(\nu m,SL^q) = Q(\nu m,SL^{p'})$ for every q with $p' \leqq q < +\infty$.

Let us now consider quadrature domains for class SL^∞. In this case $\log(1/|z-q|) \notin L^\infty(\Omega)$ if $q \in \partial\Omega$. Therefore we can not use the same argument as in Theorem 3.5. Let $[\Omega]^{cap}$ be the maximal domain of Ω with respect to capacity, namely, $[\Omega]^{cap} = \{z \in \mathbb{C}|\ cap(\Delta(r;z)\setminus\Omega) = 0 \text{ for some } r > 0\}$. Let $q \notin [\Omega]^{cap}$, let μ_ε be the equilibrium distribution of $E_\varepsilon \equiv ([\Omega]^{cap})^c \cap \overline{\Delta(\varepsilon;q)}$ and let u_ε be the conductor potential of E_ε, namely, $u_\varepsilon(z) =$

$$u_\varepsilon(z) = \int_{E_\varepsilon} \log \frac{1}{|z-\zeta|}\, d\mu_\varepsilon(\zeta).$$

Then u_ε is bounded from above and harmonic on E_ε^c. Therefore, by taking sufficiently small ε, we can replace $\max\{\frac{1}{2}+\log\frac{10\lambda(p)}{|z-q|}, 0\}$ by $\max\{\frac{1}{2}+\log 10\lambda(p)+u_\varepsilon(z), 0\}$ in the proof of Theorem 3.5.

Since a set of capacity zero is removable for functions of class SL^∞, we see that there exists the minimum domain $\tilde{W}$ in $Q(\nu m, SL^\infty)$ satisfying $[\tilde{W}]^{cap} = \tilde{W}$, namely, there is a domain $\tilde{W} = [\tilde{W}]^{cap} \in Q(\nu m, SL^\infty)$ such that $\Omega \in Q(\nu m, SL^\infty)$ if and only if $[\tilde{W}]^{cap} \subset [\Omega]^{cap}$ and $m([\Omega]^{cap} \setminus [\tilde{W}]^{cap}) = 0$. The domain $\tilde{W}$ is determined uniquely.

In what follows we shall concern with quadrature domains for classes SL^1, HL^1 and AL^1. Results on quadrature domains for classes SL^p, HL^p and AL^p with $p > 1$ are obtained by doing suitable modification.

Finally we note here that if ν is a finite positive measure with compact support and if ν is singular with respect to m, then there is an open set Ω satisfying from (1) to (3) in Introduction for class SL^1 (see Lemma 14.6).

§4. Modifications under restrictions

The main purpose of this section is to prove Theorem 4.7. From this theorem we can think of the method of construction of quadrature domains for class HL^1 and AL^1 (see §5). Further application will be given in §10. The theorem will be proved by considering modifications of measures restricted in a given domain.

Let μ be a finite positive measure on $\mathbb{C}$. For a fixed number $N > 0$, we define a function $\lambda_S(z) = \lambda_S(z;\mu,N)$ on $\mathbb{C}$ by

$$\lambda_S(z) = \sup\{r \geqq 0 \mid \mu(S(r;z)) \geqq 4Nr^2\},$$

where $S(r;z)$ denotes the closed square with horizontal and vertical sides of length $2r$ and center at z. As the function λ defined in §2, λ_S satisfies

(1) λ_S is nonnegative upper semicontinuous on $\mathbb{C}$.

(2) $\mu(S(\lambda_S(z);z)) = 4N(\lambda_S(z))^2$.

We also have Lemma 2.2 replacing $\lambda(z)$ by $\lambda_S(z)$. Here (b) in (3) of Lemma 2.2 should be replaced the following:

(b_S) There are no open squares S with horizontal and vertical sides such that $d\mu/dm = N$ a.e. on S.

The following lemma makes clear the difference between λ and λ_S.

Lemma 4.1. For every $\zeta \in \mathbb{C}$, there exist $z \in \mathbb{C}$ and a Borel measurable function g on $\mathbb{C}$ such that $0 \leqq g \leqq 1$, $\lambda_S(z;g\mu,N) = (1/2)\lambda_S(\zeta;\mu,N)$ and $g = 0$ a.e. on $S(\lambda_S(z;g\mu,N);z)^c$.

Proof. We may assume $r = \lambda_S(\zeta;\mu,N) > 0$. Divide the square $S(r;\zeta)$ into four squares by cutting it along the horizontal and vertical lines passing through ζ. Then at least one square, say S, satisfies $\mu(S) \geqq Nr^2$. Hence the center z of S and $g = (Nr^2/\mu(S))\chi_S$ have the required properties.

Using an argument similar to Lemma 3.2, we have

Lemma 4.2. Let μ be a finite positive measure with compact support K and suppose $\lambda_S(q) = \lambda_S(q;\mu,N) > 0$ on K. Set $U = \cup_{q\in K} \Delta(3\lambda_S(q);q)$. Then

$$\int_K s d\mu \leqq 46N \int_U s dm$$

for every nonnegative $s \in SL^1(U)$.

Proof. Set $\Lambda_S(p) = \sup_{q\in K} \{\lambda_S(q) - |q-p|/3\}$. Then Λ_S satisfies from (1) to (6) in the proof of Lemma 3.2, replacing λ (resp. Λ) by λ_S (resp. Λ_S). Here the constant $4N\pi$ in the inequality of (a) in (6) should be replaced by 16N. Therefore we have the lemma, by replacing the constant 36N in (3.1) by $46N > (144/\pi)N$.

By using these lemma we obtain the following propositions concerning modifications under restrictions.

Proposition 4.3. Let R be a domain in $\mathbb{C}$ and let R_0 be a domain with piecewise smooth boundary such that $\overline{R}_0 \subset R$. Let W be a bounded domain with quasi-smooth boundary such that $\overline{W} \subset R$ and $\overline{R}_0 \cap \partial W$ consists of piecewise smooth arcs. Let ν be a finite positive measure on $\overline{W}$ with $d\nu/dm \geqq \chi_W$, where $d\nu/dm$ denotes the Radon-Nikodym derivative of ν with respect to m. Let $N \geqq 72\pi$, $\beta = \beta(\nu-\chi_W m, W)$ and $\lambda_S(z) = \lambda_S(z;\chi_{\overline{R}_0}\beta,N)$.

Suppose $\inf_{z\in\overline{R}_0\cap\partial W} \lambda_S(z) > 0$. Then, for every $\varepsilon > 0$, one can construct a bounded domain W' with quasi-smooth boundary and a finite positive measure ν' on $\overline{W'}$ with $d\nu'/dm \geqq \chi_{W'}$ satisfying the following conditions:

(1) $\overline{R}_0 \cap \partial W'$ consists of piecewise smooth arcs.

(2) $W \subset W' \subset \overline{W'} \subset R$.

(3) $\int_{\overline{W}} s d\nu \leqq \int_{\overline{W'}} s d\nu'$ for every $s \in S(\overline{W'})$.

(4) $m(W') \leqq \nu(\overline{W})$.

(5) Set $\beta' = \beta(\nu'-\chi_{W'}m, W')$ and $\lambda_S'(z) = \lambda_S(z;\chi_{\overline{R}_0}\beta', N)$. Then $\inf_{z\in\overline{R}_0\cap\partial W'}\lambda_S'(z) > 0$ and $\max_{z\in\overline{R}_0\cap\partial W'}\lambda_S'(z) < \varepsilon$.

__Proof__. Assume $\lambda_S(\zeta) \geqq \varepsilon$ for some $\zeta \in \overline{R}_0 \cap \partial W$. Let $d = d(\overline{R}_0, \partial R) > 0$. If $\lambda_S(\zeta) \geqq d/10\sqrt{2}$, then, by repeating use of Lemma 4.1, we can find $z \in \mathbb{C}$ and g such that $0 \leqq g \leqq 1$ and

$$d/20\sqrt{2} \leqq \lambda_S(z;g\chi_{\overline{R}_0}\beta, N) < d/10\sqrt{2}.$$

If $\lambda_S(\zeta) < d/10\sqrt{2}$, then take $z = \zeta$ and $g \equiv 1$.

Set $r = \lambda_S(z;g\chi_{\overline{R}_0}\beta, N)$. Since $S(r;z) \subset \overline{\Delta(\sqrt{2}r;z)}$, we have

$$(g\chi_{\overline{R}_0}\beta)(\overline{\Delta(\sqrt{2}r;z)}) \geqq 4Nr^2 \geqq 144\pi(\sqrt{2}r)^2.$$

By using Lemma 2.1, choose ρ with $8\sqrt{2}r \leqq \rho \leqq 9\sqrt{2}r$ so that $W_1 = W \cup \Delta(\rho;z)$ is a domain with quasi-smooth boundary and $\overline{R}_0 \cap \partial W_1$ consists of piecewise smooth arcs.

Since there is a point $q \in (\overline{R}_0 \cap \partial W) \cap S(r;z)$, $|p-z| \geqq |p-q| - |q-z| \geqq d - \sqrt{2}r > 9\sqrt{2}r$ for $p \notin R$ and so $\overline{W_1} \subset R$. Set

$$\nu_1 = \chi_W m + \chi_{(\partial W)\setminus\overline{R}_0}\beta + (1-g)\chi_{\overline{R}_0}\beta + fm,$$

where $f(\cdot) = f(\cdot;g\chi_{\overline{R}_0}\beta,\Delta(\rho;z))$. Then $m(W_1) \leqq \nu_1(\overline{W_1}) = \nu(\overline{W})$ and $\int_{\overline{W}} s d\nu \leqq \int_{\overline{W_1}} s d\nu_1$ for every $s \in S(\overline{W_1})$. Let $\beta_1 = \beta(\nu_1 - \chi_{W_1} m, W_1)$ and $\lambda_{S,1}(p) = \lambda_S(p;\chi_{\overline{R}_0}\beta_1, N)$. Then $\inf_{p\in\overline{R}_0\cap\partial W_1}\lambda_{S,1}(p) > 0$, since W_1 is a domain with quasi-smooth boundary and $\overline{R}_0 \cap \partial W_1$ consists of piecewise smooth arcs.

If $\max_{p\in\overline{R}_0\cap\partial W_1}\lambda_{S,1}(p) \geqq \varepsilon$, then we again repeat the above process. Since $r \geqq \min\{\varepsilon, d/20\sqrt{2}\} > 0$, our process must stop after a finite number of times and yields a domain W' and a measure ν' satisfying from (1) to (5).

<u>Proposition 4.4</u>. Further if $\max_{z\in\overline{R}_0\cap\partial W}\lambda_S(z) < d(\overline{R}_0,\partial R)/10\sqrt{2}$ in Proposition 4.3, then one can construct W' and ν' satisfying the following additional condition:

(6) $U \subset W'$, $\overline{U'} \subset R$ and $U \cap U' = \phi$, where $U = \cup_{q\in\overline{R}_0\cap\partial W}\Delta(3\lambda_S(q);q)$ and $U' = \cup_{q\in\overline{R}_0\cap\partial W'}\Delta(3\lambda_S'(q);q)$.

<u>Proof</u>. We recall the argument used at the beginning of the proof of Proposition 3.3. Set $E_1 = \overline{R}_0 \cap W$ and $E_{j+1} = E_j \setminus \Delta(2\sqrt{2}\lambda_S(p_j);p_j)$, $j = 1,\cdots,n$, as in the proof of Proposition 3.3 and choose r_j with $8\sqrt{2}\lambda_S(p_j) \leqq r_j \leqq 9\sqrt{2}\lambda_S(p_j)$ so that $W_0 = W \cup \cup_{j=1}^{n}\Delta(r_j;p_j)$ is a domain with quasi-smooth boundary and $\overline{R}_0 \cap \partial W_0$ consists of piecewise smooth arcs. Since $\max_{z\in\overline{R}_0\cap\partial W}\lambda_S(z) < d(\overline{R}_0,\partial R)/10\sqrt{2}$, $r_j < d(\overline{R}_0,\partial R)$ and so $\overline{W}_0 \subset R$. It is easy to show that $U \subset W_0$.

Apply Proposition 4.3 to W_0 and ν_0 for $\varepsilon = \min\{d(\overline{R}_0,\partial R)/10\sqrt{2}, \inf_{q\in\overline{R}_0\cap\partial W}\lambda_S(q)\}$. By using an argument similar to Proposition

3.3, we see that U and U' satisfy (6).

Proposition 4.5. Let R be a domain in $\mathbb{C}$, let W be a bounded domain with quasi-smooth boundary such that $\overline{W} \subset R$ and let ν be a finite positive measure on $\overline{W}$ with $d\nu/dm \geqq \chi_W$ and $\int d\nu > m(W)$. Then one can construct a sequence $\{W_R^{(n)}\}$ of bounded domains with quasi-smooth boundaries and a sequence $\{\nu_R^{(n)}\}$ of finite positive measures satisfying the following:

(1) $\overline{W} \subset W_R^{(n)} \subset \overline{W_R^{(n)}} \subset W_R^{(n+1)} \subset G$ for every n, where G denotes the connected component of $R \cap \tilde{W}$ containing $\overline{W}$. The domain $\tilde{W}$ is the minimum in $Q(\nu, SL^1)$ constructed in Theorems 3.4 and 3.5.

(2) $\operatorname{supp} \nu_R^{(n)} = \overline{W_R^{(n)}}$ and $d\nu_R^{(n)}/dm \geqq \chi_{W_R^{(n)}}$.

(3) $\int s\,d\nu \leqq \int s\,d\nu_R^{(n)}$ for every $S(\overline{W_R^{(n)}})$.

(4) $\beta_R^{(n)}(K) \to 0$ as $n \to \infty$ for every compact subset K of R, where $\beta_R^{(n)} = \beta(\nu_R^{(n)} - \chi_{W_R^{(n)}} m, W_R^{(n)})$.

Proof. For every $\delta > 0$, we set $R_\delta = \{z \in R \mid d(z,\partial R) > \delta$ and $|z| < 1/\delta\}$. Take $\delta_0 > 0$ so small that $\overline{W} \subset R_{\delta_0}$ and let R_0 be a domain with piecewise smooth boundary such that $R_{\delta_0} \subset R_0 \subset \overline{R}_0 \subset R$. Since W is a domain with quasi-smooth boundary and $\overline{R}_0 \cap \partial W = \partial W$, by Proposition 2.5, we have $\inf_{z\in\partial W}\lambda_S(z) > 0$. Applying Propositions 4.3 and 4.4 to W and ν for $\varepsilon = d(\overline{R}_0,\partial R)/10\sqrt{2}$, we obtain W' and ν' satisfying from (1) to (6) in Propositions 4.3 and 4.4.

Next take $\delta_1 > 0$ so that $\delta_1 < \delta_0$ and $\overline{W'} \subset R_{\delta_1}$ and let R_1 be a domain with piecewise smooth boundary such that $R_{\delta_1} \subset R_1 \subset \overline{R_1} \subset R$. Apply Propositions 4.3 and 4.4 to W' and ν' for $\varepsilon = d(\overline{R_1}, \partial R)/10\sqrt{2}$ replacing R_0 by R_1 and we obtain $W^{(2)}$ and $\nu^{(2)}$.

Take $\{\delta_n\}$ so that $\delta_n \downarrow 0$ as $n \uparrow \infty$, repeat this process and set $W_R^{(n)} = W^{(n)} = (W^{(n-1)})'$ and $\nu_R^{(n)} = \nu^{(n)} = (\nu^{(n-1)})'$. Then it is easy to show that these satisfy from (1) to (3). Let K be a compact subset of R. Then $K \subset R_k$ for some k. We may assume that $\overline{R}_k \cap \partial W^{(n)}$ consists of piecewise smooth arcs for every $n \geqq k$. Hence, by Lemma 4.2,

$$\int_{\overline{R}_k \cap \partial W_R^{(n)}} d\beta_R^{(n)} \leq 46N \int_{U_k^{(n)}} dm$$

for every $n \geqq k$, where $U_k^{(n)} = \cup_{q \in \overline{R}_k \cap \partial W_R^{(n)}} \Delta(3\lambda_S(q;\chi_{\overline{R}_k}\beta_R^{(n)}, N); q)$. Since, for every fixed k, $\{U_k^{(n)}\}_{n=k}^{\infty}$ are mutually disjoint and $U_k^{(n)} \subset W^{(n+1)} \subset \tilde{W}$, $m(U_k^{(n)}) \to 0$ as $n \to \infty$. Hence (4) holds.

Let $\{W_R^{(n)}\}$ and $\{\nu_R^{(n)}\}$ be sequences of domains and measures constructed in Proposition 4.5 by taking $N \geqq 50e\pi$ and fix them. Set $\tilde{W}_R = \cup W_R^{(n)}$. Then we have

<u>Proposition 4.6</u>. It follows that

(1) $\tilde{W} \subset R$ if and only if $\tilde{W}_R = \tilde{W}$.

(2) If $\tilde{W} \subset R$, then $(\nu_R^{(n)} - \chi_{W_R^{(n)}} m)(\overline{W_R^{(n)}}) \to 0$ as $n \to \infty$.

(3) If $(\nu_R^{(n)} - \chi_{W_R^{(n)}} m)(\overline{W_R^{(n)}}) \to 0$ as $n \to \infty$, then $[\tilde{W}]^{cap} = [\tilde{W}_R]^{cap}$.

<u>Proof</u>. It is easy to show that (1) and (2) hold. Assume

$(\nu_R^{(n)} - \chi_{W_R^{(n)}} m)(\overline{W_R^{(n)}}) \to 0$ as $n \to \infty$. Then $\tilde{W}_R \in Q(\nu, SL^\infty)$ and so $[\tilde{W}]^{cap} = [\tilde{W}_R]^{cap}$ by the argument given at the end of §3.

Let W and ν be as in Theorem 3.4. In Theorem 3.5, we have proved that there exists the minimum domain $\tilde{W}$ in $Q(\nu, SL^1)$. Let us show a remarkable relation between $\Omega \in Q(\nu, HL^1)$ and $\tilde{W}$.

Theorem 4.7. Let Ω be a domain in $Q(\nu, HL^1) \backslash Q(\nu, SL^1)$ containing $\overline{W}$. Let ω be the harmonic measure on $\partial\Omega$ with respect to a fixed point $a \in \Omega$. Then

$$\omega(\overline{\Delta(r;p)} \cap \partial\Omega) < c\pi r^2$$

for every $r > 0$ and every $p \in (\partial\Omega) \backslash \tilde{W}$, where c is a constant independent of r and p. If $p \in (\partial\Omega) \cap \tilde{W}^e$, then $\omega(\overline{\Delta(r;p)} \cap \partial\Omega) = 0$ for some $r > 0$.

Proof. Apply Proposition 4.5 replacing R by Ω. Let $\{W_\Omega^{(n)}\}$ and $\{\nu_\Omega^{(n)}\}$ be sequences of domains and measures satisfying from (1) to (4) in Proposition 4.5, respectively.

Let $p \in (\partial\Omega) \backslash \tilde{W}$ and $\beta_\Omega^{(n)} = \beta(\nu_\Omega^{(n)} - \chi_{W_\Omega^{(n)}} m, W_\Omega^{(n)})$. If $\lambda(p; \beta_\Omega^{(n)}, 144) > 0$, then, by using an argument similar to Proposition 3.3, we see that $p \in \tilde{W}$. This is a contradiction. Hence $\lambda(p; \beta_\Omega^{(n)}, 144) = 0$ for every n, namely, $\beta_\Omega^{(n)}(\overline{\Delta(r;p)}) < 144\pi r^2$ for every $r > 0$ and n.

Let $\omega(z; \overline{\Delta(r;p)} \cap \partial\Omega, \Omega)$ be the solution in Ω of the Dirichlet problem for the boundary function $\chi_{\overline{\Delta(r;p)} \cap \partial\Omega}$. Then, by definition, $\omega(\overline{\Delta(r;p)}\cap\partial\Omega) = \omega(a;\overline{\Delta(r;p)}\cap\partial\Omega,\Omega)$.

Set $\omega(r) = \omega(\overline{\Delta(r;p)}\cap\partial\Omega)$ and assume $\omega(r) > 0$. Let $h(z) = \omega(z;\overline{\Delta(r;p)}\cap\partial\Omega,\Omega)/\omega(r)$. Then $h \in HL^\infty(\Omega) \subset HL^1(\Omega)$ and $0 \leqq h \leqq 1/\omega(r)$ on Ω. Since $\Omega\setminus W_\Omega^{(n)} \supset \Omega\setminus\tilde{W}_\Omega \supset \Omega\setminus\tilde{W}$, $m(\Omega\setminus\tilde{W}) > 0$ and $h(a) = 1$, by using the Harnack inequality, we can find a constant $k > 0$ independent of p, r and n such that

$$\int h d\beta_\Omega^{(n)} = \int_{\Omega\setminus W_\Omega^{(n)}} h dm \geqq k.$$

Since

$$\int h d\beta_\Omega^{(n)} \leqq \frac{\beta_\Omega^{(n)}(\overline{\Delta(R;p)})}{\omega(r)} + \int_{\Omega\setminus\overline{\Delta(R;p)}} h d\beta_\Omega^{(n)}$$

$$\leqq \frac{144\pi R^2}{\omega(r)} + \int_{\Omega\setminus\Delta(R;p)} h d\beta_\Omega^{(n)}$$

for every $R > r$, taking $c > 144/k$ we have the required inequality if we show

$$\lim_{n\to\infty} \int_{\Omega\setminus\Delta(R;p)} h d\beta_\Omega^{(n)} = 0.$$

Let u be the Evans-Selberg function with respect to the set I of all irregular points on $\partial\Omega$, namely, let u be the logarithmic potential of a finite positive measure concentrated on I such that u(z) tends to $+\infty$ when z tends to any point of I. Since Ω is bounded (this will be proved in Theorem 6.4), we can find $\alpha > 0$ so that $v(z) = u(z) + \alpha$ is positive on Ω.

For every $\varepsilon > 0$, let $U_\varepsilon = \{z \in \Omega | h(z) < \varepsilon\}$ and $V_\varepsilon = \{z \in \Omega | v(z) > 1/\varepsilon\}$. Since $v \in HL^1(\Omega)$,

$$\beta_\Omega^{(n)}(V_\varepsilon)/\varepsilon \leqq \int v d\beta_\Omega^{(n)} = \int_{\Omega\setminus W_\Omega^{(n)}} vdm.$$

Hence

$$\int_{V_\varepsilon} hd\beta_\Omega^{(n)} \leqq \frac{\int_\Omega vdm}{\omega(r)}\varepsilon.$$

Since

$$\int_{U_\varepsilon} hd\beta_\Omega^{(n)} \leqq \beta_\Omega^{(n)}(U_\varepsilon)\varepsilon \leqq m(\Omega)\varepsilon$$

and $\Omega\setminus\Delta(R;p)\setminus U_\varepsilon\setminus V_\varepsilon$ is a compact subset of Ω, by (4) of Proposition 4.5,

$$\limsup_{n\to\infty} \int_{\Omega\setminus\Delta(R;p)} hd\beta_\Omega^{(n)} \leqq \left(\frac{\int_\Omega vdm}{\omega(r)} + m(\Omega)\right)\varepsilon.$$

Hence

$$\lim_{n\to\infty} \int_{\Omega\setminus\Delta(R;p)} hd\beta_\Omega^{(n)} = 0.$$

Finally let $p \in (\partial\Omega) \cap \tilde{W}^e$. Then $\beta_\Omega^{(n)}(\overline{\Delta(r;p)}) = 0$, $n = 1, 2, \cdots$, for some $r > 0$. If $\omega(r) > 0$, then, by the above argument, we have $k = 0$. This is a contradiction. Hence $\omega(r) = 0$.

Theorem 4.7 gives the property of quadrature domains for class HL^1. The same does not hold for class AL^1 in general. We can see it from Example 1.2.

Corollary 4.8. Let Ω be as in Theorem 4.7. If Ω is a domain with quasi-smooth boundary, then $\partial\Omega \subset \tilde{W}$.

Proof. Let δ_a be the Dirac measure at a. Then $\omega = \beta(\delta_a,\Omega)$

and so $\inf_{z\in\partial\Omega}\lambda(z;\omega,c) > 0$, by Proposition 2.5. Hence, by Theorem 4.7, $(\partial\Omega)\setminus\tilde{W} = \phi$, namely, $\partial\Omega \subset \tilde{W}$.

Corollary 4.9. Let Ω be as in Theorem 4.7. Then $\omega((\partial\Omega)\setminus\tilde{W}) = 0$ and $\mathrm{cap}((\partial\Omega)\cap\tilde{W}^e) = 0$.

Proof. By Theorem 4.7, $\omega((\partial\Omega)\cap\tilde{W}^e) = 0$. Since $\tilde{W}^e$ is open, $\mathrm{cap}((\partial\Omega)\cap\tilde{W}^e) = 0$. Set $\mu = \omega|(\partial\Omega)\cap(\partial\tilde{W})$. Then $\lambda(z;\mu,c) \leqq \lambda(z;\omega,c) = 0$ on $(\partial\Omega)\cap(\partial\tilde{W})$. By Lemma 2.2, $\mu << m$. We shall see in Corollary 10.10 that $m(\partial\tilde{W}) = 0$. Hence $\mu = 0$ so that $\omega((\partial\Omega)\setminus\tilde{W}) = \omega((\partial\Omega)\cap\tilde{W}^e) + \omega((\partial\Omega)\cap(\partial\tilde{W})) = 0$.

In Theorem 6.4, we shall show that $\tilde{W}$ is bounded. Hence the unbounded component of $\tilde{W}^e$ is determined uniquely.

Corollary 4.10. Let Ω be a domain in $Q(\nu,HL^1)$ containing $\overline{W}$. Then $\overline{\Omega}$ is contained in the complement E of the unbounded component of $\tilde{W}^e$.

Proof. We may assume $\Omega \notin Q(\nu,SL^1)$. Then, by Corollary 4.9, $(\partial\Omega) \cap E^c = \phi$. Hence $\partial\Omega \subset E$ so that $\overline{\Omega} \subset E$.

§5. Construction of quadrature domains for harmonic and analytic functions

In this section we consider the method of construction of quadrature domains. We shall define three operations S, H and A. Starting from a domain W with quasi-smooth boundary and a finite positive measure ν on W satisfying $d\nu/dm \geqq \chi_W$ and $\int d\nu >$

m(W), we use these operations sequentially and obtain the quadrature domain of ν.

Let us first recall Proposition 3.3. In the proposition we have constructed new W' and ν' satisfying from (1) to (4). The assumption $\inf_{z\in\partial W}\lambda(z) > 0$ is inessential, because we can apply the argument used in the beginning of the proof of Theorem 3.4. We call the process constructing W' and ν' from W and ν the operation S and denote $\{W',\nu'\} = S\{W,\nu\}$. We note here that $\{W',\nu'\}$ is not determined uniquely.

Next let us define the operation H. Let O be the union of a finite number of bounded connected components of the exterior of W and set $W' = (\overline{W\cup O})^{\circ}$. Then W' is a domain with quasi-smooth boundary containing W. If $\beta(\nu-\chi_{W'}m,W') = \beta(\nu-\chi_W m,W') - \beta(\chi_O m,W')$ is nonnegative, we set $\nu' = \chi_{W'}m + \beta(\nu-\chi_{W'}m,W')$. Then $\int h d\nu = \int h d\nu'$ for every harmonic function on a neighborhood of $\overline{W'}$. We denote $\{W',\nu'\} = H\{W,\nu\}$ and say $\{W',\nu'\}$ is constructed from $\{W,\nu\}$ by the operation H. We note that $\{W',\nu'\}$ is not determined uniquely and there is a possibility that $\nu' = \chi_{W'}m$.

Finally let us define the operation A. Let $W' = (\overline{W})^{\circ}$ and let μ be a finite real measure on $\partial W'$ satisfying $\int f d\mu = 0$ for every f in the class AC(W') of functions analytic on W' and continuous on $\overline{W'}$. If $\beta(\nu-\chi_W m,W') - \mu$ is nonnegative, we set $\nu' = \chi_{W'}m + \beta(\nu-\chi_W m,W') - \mu$, denote $\{W',\nu'\} = A\{W,\nu\}$ and say $\{W',\nu'\}$ is constructed from $\{W,\nu\}$ by the operation A.

Here we note that there are many such a measure μ. Let W be a domain with quasi-smooth boundary satisfying $W = (\overline{W})^{\circ}$ and let V(W) be the real linear space of finite real measures μ on ∂W satisfying $\int f d\mu = 0$ for every $f \in AC(W)$. Then we have

Lemma 5.1. If W is n-ply connected, then $\dim_{\mathbb{R}} V(W) = n - 1$.

Proof. First let us show that $\dim_{\mathbb{R}} V(W) \leqq n - 1$. Assume that ∂W consists of mutually disjoint smooth simple curves Γ_0, $\Gamma_1, \cdots, \Gamma_{n-1}$. Let ω_j, $j = 1, 2, \cdots, n - 1$, be the harmonic measure of Γ_j, namely, the harmonic function on W which is continuously extensible to ∂W satisfying $\omega_j = 1$ on Γ_j and $\omega_j = 0$ on $(\partial W)\setminus\Gamma_j$ and let w_j be the mutiple-valued analytic function on W such that $\mathrm{Re}\, w_j = \omega_j$.

For a real valued function u of class C^1 on ∂W, we can also consider the multiple-valued analytic function f such that $\mathrm{Re}\, f = u$. It is well-known that one can find real constants a_j, $j = 1, 2, \cdots, n - 1$, so that $f - \Sigma a_j w_j \in AC(W)$.

If $\mu_1, \mu_2 \in V(W)$ satisfy $\int_{\Gamma_j} d\mu_1 = \int_{\Gamma_j} d\mu_2$, $j = 1, 2, \cdots, n - 1$, then

$$\int u d(\mu_1 - \mu_2) = \mathrm{Re} \int (f - \sum a_j w_j) d(\mu_1 - \mu_2) = 0.$$

This implies $\mu_1 = \mu_2$ and so $\dim_{\mathbb{R}} V(W) \leqq n - 1$.

Now consider the case that W is a domain with quasi-smooth boundary satisfying $W = (\overline{W})^{\circ}$. Let E be the set of ends for some expression of ∂W. By using an argument similar to the above, we see that if $\mu_1, \mu_2 \in V(W)$ satisfy $\int_{\Gamma_j} d\mu_1 = \int_{\Gamma_j} d\mu_2$, then $\mu_1 = \mu_2$ on $(\partial W)\setminus E$. Since E is finite, it follows that $\mu_1 = \mu_2$.

Next we show that there exist measures $\mu_k \in V(W)$, $k = 1, 2, \cdots, n - 1$, such that $\int_{\Gamma_j} d\mu_k = \delta_{jk}$, where δ_{jk} denotes the Kronecker delta. We may assume that ∂W consists of mutually disjoint smooth simple curves $\Gamma_0, \Gamma_1, \cdots, \Gamma_{n-1}$. Let $\zeta_k = \varphi_k(z)$ be a one-to-one conformal mapping from W onto a circular slit annulus R_k with center at the origin corresponding Γ_0 to the inner circle and Γ_k to the outer circle and let μ_k be the measure on ∂W defined by $d\mu_k = (1/(2\pi))d \arg \varphi_k$. Then μ_k is the desired measure, because

$$\int_{\partial W} f d \arg \varphi_k = \int_{\partial R_k} \frac{f(\varphi_k^{-1}(\zeta_k))}{i\zeta_k} d\zeta_k = 0$$

for every $f \in AC(W)$. This completes the proof.

Let us consider next the composition of operations. The composition means the composition of an infinite number of operations containing an infinite number of operations S except the composition of a finite number of operations whose last operation is H and the final measure is equal to $\chi_{W_0} m$, where W_0 denotes the final domain. Now we summarize our method of construction of quadrature domains.

<u>Theorem 5.2</u>. Let C be the composition of operations S, H and A, and let C_n be its n-th partial composition. Set $\{W^{(n)}, \nu^{(n)}\} = C_n\{W,\nu\}$. Then

(1) $\cup W^{(n)} \in Q(\nu, AL^1)$.

(2) $\cup W^{(n)} \in Q(\nu, HL^1)$ if C consists of operations S and H.

(3) $\cup W^{(n)} \in Q(\nu, SL^1)$ if C consists of operations S only.

Proof. If C is the composition of an infinite number of operations, then the theorem follows from an argument similar to Theorem 3.4. If C is the composition of a finite number of operations, by definition, the final $W^{(n)}$ is a domain with quasi-smooth boundary satisfying $(\overline{W^{(n)}})^\circ = W^{(n)}$. Since the class of functions harmonic on $\overline{W^{(n)}}$ is dense in $HL^1(W^{(n)})$ (see Lemma 7.3), we also have (2) in this case.

CHAPTER II. PROPERTIES OF QUADRATURE DOMAINS

§6. Basic properties of quadrature domains

In this section we shall consider properties of quadrature domains. Throughout this section we assume that the support supp ν of a measure ν is compact. Basic properties of quadrature domains are

(1) Domains in $Q(\nu, AL^1)$ are uniformly bounded.

(2) If $\Omega \in Q(\nu, AL^1)$ and supp $\nu \subset \Omega$, then every isolated nondegenerate boundary component is an analytic quasi-simple curve.

(3) If $\Omega \in Q(\nu, AL^1)$ and supp $\nu \subset \Omega$, then $[\Omega]\backslash\Omega$ is contained in a real analytic set in $[\Omega]\backslash$supp ν.

First we prepare two lemmas. Lemma 6.1 is a kind of comparison theorem.

Lemma 6.1. Let $v(r)$ be a nonnegative measurable function on $[0,+\infty)$ satisfying

$$\int_r^\infty v(t)dt \leqq v(r)^2\{A+B \log \frac{1}{v(r)}\}$$

for every r with $0 \leqq v(r) \leqq 1/e$, where the right-hand side is equal to 0 if $v(r) = 0$. Then v vanishes almost everywhere on $[e\int_0^\infty vdt+(2A+3B)/e,+\infty)$.

Proof. Set $x(y) = (2A+3B)/e - (2A+B)y - 2B\, y \log(1/y)$. Since $y \log(1/y)$ is increasing on $(0,1/e]$, $x(y)$ decreases from $(2A+3B)/e$ to 0 when y increases from 0 to $1/e$. Let $y = y(x)$ be the inverse function of $x(y)$ on $[0,(2A+3B)/e)$ and let $y = 0$ on $[(2A+3B)/e,+\infty)$. Then $y(0) = 1/e$, y is decreasing and

$$\int_r^\infty y(t)dt = y(r)^2\{A+B \log \frac{1}{y(r)}\}$$

for every $r \in [0,(2A+3B)/e)$.

Assume that v satisfies $v(r) \leqq 1/e$ and the above inequality, and let $Y(r) = \int_r^\infty y(t)dt$ and $V(r) = \int_r^\infty v(t)dt$. Then $Y(r) < V(r)$ for some $r \in [0,(2A+3B)/e)$ implies $y(r) < v(r)$. Since Y and V are both continuous, this implies $Y(t) < V(t)$ and $y(t) < v(t)$ for every $t \leqq r$. Hence $1/e = y(0) < v(0) \leqq 1/e$. This contradiction implies that $V(r) \leqq Y(r)$ for every $r \in [0,(2A+3B)/e)$. Since $Y((2A+3B)/e) = 0$, v vanishes almost everywhere on $[(2A+3B)/e,+\infty)$. The lemma follows from the fact that $m_1\{r \geqq 0 |\ v(r) > 1/e\}$ is not greater than $e\int_0^\infty vdt$.

Lemma 6.2. The inequality

$$\frac{1}{2\pi}\int_{\Delta(1;0)} \frac{dm(w)}{|w-z|\,|w+z|} < 1 + \log \frac{1}{|z|}$$

holds for every $z \in \overline{\Delta(1;0)}\setminus\{0\}$.

Proof. Let us write $s(z)$ the value of the left-hand side of the above inequality. Then s is subharmonic on $\Delta(1;0)\setminus\{0\}$, is continuous on $\overline{\Delta(1;0)}\setminus\{0\}$ and satisfies $s(z) = s(|z|)$. Since $s(z) - \log(1/|z|)$ is bounded in a neighborhood of 0, the lemma follows from the following estimation:

$$2\pi s(1) = 2\int_{\Delta(1;0)\cap\{\mathrm{Re}\, w>0\}} \frac{dm(w)}{|w-1||w+1|}$$

$$< 2\int_{\Delta(1;0)\cap\{\mathrm{Re}\, w>0\}} \frac{dm(w)}{|w-1|}$$

$$< 2\int_{\Delta(1;1)\cap\{\mathrm{Re}\, w<1\}} \frac{dm(w)}{|w-1|}$$

$$= 2\pi.$$

Now let us define a one-parameter family of domains.

Let D be a bounded domain whose boundary consists of a finite number of mutually disjoint analytic simple curves. Let G_j, $j = 0, 1, \cdots, n-1$, be connected components of D^e; we assume $\infty \in G_0$. Let φ_j be a one-to-one conformal mapping from $\overline{G}_j$ onto $\overline{\Delta(1;0)}$ (onto $\Delta(1;0)^c$ such that $\varphi_j(\infty) = \infty$ if $j = 0$) and set $\gamma_j = \max_{z\in\partial G_j} |\varphi_j'(z)|/\min_{z\in\partial G_j} |\varphi_j'(z)|$ and $\gamma = \max_{0\leqq j\leqq n-1} \{\gamma_j\}$.

We consider arcs $L_j = \varphi_j^{-1}([0,1])$ $(= \varphi_j^{-1}([1,+\infty))$ if $j = 0)$ and define a function $r(z)$ on D^c as follows: If $z \in L_j$, $r(z)$ is the length of the subarc of L_j starting from $\varphi_j^{-1}(1)$ and ending at z. If $z \in \overline{G}_j\setminus L_j$, then there exists one and only one curve of the form $\varphi_j^{-1}(\{|w| = \text{const.}\})$ passing through z. Let $\pi(z)$ be the crossing point of the curve and L_j, and set $r(z) = r(\pi(z))$.

For $r \geqq 0$, set $D(r) = D \cup \{z \in D^c |\ r(z) < r\}$. If necessary, we write $D(r) = D(r;D,\varphi_0,\varphi_1,\cdots,\varphi_{n-1})$. For every $r \geqq 0$, $D(r)$ is a bounded domain whose boundary consists of a finite number of mutually disjoint analytic simple curves or points.

By the Fejér-Riesz inequality, $\ell(L_j) < \frac{1}{2}\ell(\partial G_j)$ for $j \neq 0$. Hence $r(z) < \frac{1}{2}\ell(\partial G_j)$ on $\overline{G}_j$ for $j \neq 0$ so that, for every $j \neq 0$, $\overline{G}_j \subset D(r)$ if $r \geqq \frac{1}{2}\ell(\partial G_j)$.

Let ν be a complex measure whose support is contained in $\overline{D}$ and set $\|\nu\| = \int d|\nu|$. Let us define $\delta(\sqrt{\|\nu\|})$ by

$$\delta(\sqrt{\|\nu\|}) = \inf\{d \geqq 0;\ D \oplus \sqrt{\|\nu\|} \subset D(d)\},$$

where $D \oplus \sqrt{\|\nu\|} = \{z \in \mathbb{C} \mid d(z,D) < \sqrt{\|\nu\|}\}$. The next lemma is easily verified.

Lemma 6.3. It follows that

(1) $\gamma \geqq 1$ and the equality holds if and only if ∂G_j is a circle $\partial\Delta(\rho_j;c_j)$ and $\varphi_j(z) = (e^{i\theta_j}/\rho_j)(z-c_j)$ for every j, where θ_j is a real constant.

(2) $\delta(\sqrt{\|\nu\|}) \geqq \sqrt{\|\nu\|}$ and if ∂G_j is a circle $\partial\Delta(\rho_j;c_j)$ and $\varphi_j(z) = (e^{i\theta_j}/\rho_j)(z-c_j)$ for every j, then the equality holds.

By using this notation we obtain

Theorem 6.4. If supp $\nu \subset \overline{D}$, then $\Omega \subset D(\delta(\sqrt{\|\nu\|}) + (1/(2e) + e\, 1.56\gamma)\sqrt{\|\nu\|})$ for every $\Omega \in Q(\nu, AL^1)$. Consequently, domains in $Q(\nu, AL^1)$ are uniformly bounded for every complex measure ν with compact support. The term $(1/(2e))\sqrt{\|\nu\|}$ can be eliminated if D is simply connected.

Proof. If $\overline{G}_j \subset D(r)$, then $D(r) = D(r; D \cup \overline{G}_j, \varphi_0, \cdots, \varphi_{j-1}, \varphi_{j+1}, \cdots, \varphi_{n-1})$. Hence we may assume that $G_j \backslash D(\delta(\sqrt{\|\nu\|}) + (1/(2e) +$

$e+1.56\gamma)\sqrt{\|\nu\|}) \neq \phi$ for every j. By considering a measure $(1/\|\nu\|)(\nu\circ\varphi^{-1})$ and $D(r;\varphi(D),\varphi_0\circ\varphi^{-1},\cdots,\varphi_{n-1}\circ\varphi^{-1})$, where $\varphi(z) = z/\sqrt{\|\nu\|}$, we may also assume $\|\nu\| = 1$.

Let $\Omega \in Q(\nu, AL^1)$ and $v(r) = \int_{\partial D(r)} \chi_\Omega(z)ds$, where ds denotes the line element of $\partial D(r)$. We shall apply Lemma 6.1 to this function $v(r)$.

It is easy to show that

$$\frac{1}{2\pi i}\int_{\partial D(r)} \hat{\nu}(z)dz = -\int d\nu$$

and

$$\frac{1}{2\pi i}\int_{\partial D(r)} \hat{\chi}_\Omega(z)dz = m(\Omega\setminus D(r)) - m(\Omega)$$

$$= m(\Omega\setminus D(r)) - \int d\nu$$

for $r > 0$. Hence

$$m(\Omega\setminus D(r)) = \frac{1}{2\pi i}\int_{\partial D(r)} (\hat{\chi}_\Omega - \hat{\nu})(z)\chi_\Omega(z)dz$$

$$\leqq \frac{1}{2\pi}\int_{\partial D(r)} |\hat{\chi}_\Omega(z) - \hat{\nu}(z)|\chi_\Omega(z)ds,$$

since $\hat{\chi}_\Omega(z) = \hat{\nu}(z)$ outside Ω.

Next consider $(dr/dn)(z)$, where $z \in \partial D(r)$ and dr/dn denotes the outer normal derivative of r. Then $(dr/dn)(z) = |\varphi_j'(z)|/|\varphi_j'(\pi(z))|$ if $z \in G_j$ so that $(dr/dn)(z) \leqq \gamma_j$ on G_j. Hence $(dr/dn)(z) \leqq \gamma$ on D^c and

$$m(\Omega\setminus D(r)) = \iint_{D(r)^c} \chi_\Omega(z)dnds$$

$$= \iint_{D(r)^c} \chi_\Omega(z) \frac{1}{\frac{dr}{dn}(z)} \, drds$$

$$\geq \frac{1}{\gamma}\int_r^\infty \left\{\int_{\partial D(r)} \chi_\Omega(z)\,ds\right\}dr$$

$$= \frac{1}{\gamma}\int_r^\infty v(t)\,dt.$$

We take r with $\delta(1) \leqq r \leqq \delta(1) + (e+1.56\gamma)$. Since $G_j \backslash D(\delta(1) + (1/(2e) + e + 1.56\gamma)) \neq \phi$, by the Fejér-Riesz inequality, $\ell(G_j \cap \partial D(r)) > 2\cdot 1/(2e) = 1/e$ for every $j \neq 0$. On the other hand, $\ell(G_0 \cap \partial D(r)) \geqq \ell(\partial\Delta(1;0)) = 2\pi > 1/e$. Hence, if $v(r) \leqq 1/e$, then, for every $z \in \Omega \cap \partial D(r)$, one can find $\zeta = \zeta(z) \in \Omega^c \cap \partial D(r)$ satisfying

$$\left|\int_z^\zeta ds\right| = \min_{w\in\Omega^c\cap\partial D(r)} \left|\int_z^w ds\right|,$$

where the integrals are taken along $\partial D(r)$. By definition, ζ satisfies $|\int_z^\zeta ds| \leqq v(r)/2$ and $|\zeta - z| \leqq v(r)/2$.

Since

$$\hat\chi_\Omega(z) - \hat\chi_\Omega(\zeta) = (z-\zeta)\int_{\Omega\cap\Delta} \frac{dm(w)}{(w-z)(w-\zeta)} + \int_\zeta^z \left\{\int_{\Omega\backslash\Delta} \frac{dm(w)}{(w-\eta)^2}\right\}d\eta,$$

where $\Delta = \Delta(1;(z+\zeta)/2)$ and η is on the line segment whose ends are z and ζ, by Lemma 6.2, we have

$$\frac{1}{2\pi}|\hat\chi_\Omega(z) - \hat\chi_\Omega(\zeta)| \leqq |z-\zeta|\left\{1 + \log\frac{1}{\frac{|z-\zeta|}{2}} + \frac{1}{2\pi}\frac{1}{\left(1 - \frac{|z-\zeta|}{2}\right)^2}\right\}$$

$$\leqq \frac{v(r)}{2}\left\{1 + \log 4 + \frac{1}{2\pi}\frac{1}{\left(1-\frac{1}{4e}\right)^2} + \log\frac{1}{v(r)}\right\}.$$

By the definition of $\delta(1)$, $d(\text{supp}\ \nu, \partial D(r)) \geqq 1$ so that

$$|\hat{\nu}(\zeta) - \hat{\nu}(z)| = \left|\int_z^\zeta \left\{\frac{d\nu(w)}{(w-\eta)^2}\right\} d\eta\right| \leqq \frac{v(r)}{2} ,$$

where the integral is taken along $\partial D(r)$.

Hence

$$\frac{1}{2\pi}\int_{\partial D(r)} |\hat{\chi}_\Omega(z) - \hat{\nu}(z)| \chi_\Omega(z) ds$$

$$\leqq \frac{1}{2\pi}\int_{\partial D(r)} |\hat{\chi}_\Omega(z) - \hat{\chi}_\Omega(\zeta)| \chi_\Omega(z) ds + \frac{1}{2\pi}\int_{\partial D(r)} |\hat{\nu}(\zeta) - \hat{\nu}(z)| \chi_\Omega(z) ds$$

$$\leqq \frac{v(r)^2}{2}\left[\left\{1 + \log 4 + \frac{1}{2\pi}\left(1 + \frac{1}{(1-\frac{1}{4e})^2}\right)\right\} + \log\frac{1}{v(r)}\right]$$

$$\leqq v(r)^2\left\{1.37 + 0.5 \log\frac{1}{v(r)}\right\}$$

so that

$$\int_r^\infty v(t)dt \leqq v(r)^2\left\{1.37\gamma + 0.5\gamma \log\frac{1}{v(r)}\right\}$$

for every r satisfying $\delta(1) \leqq r \leqq \delta(1) + (e+1.56\gamma)$ and $0 \leqq v(r) \leqq 1/e$. Since $e\int_{\delta(1)}^\infty v dt + (2\times 1.37 + 3\times 0.5)\gamma/e < e + 1.56\gamma$, by Lemma 6.1, $v(r) = 0$ for every $r \geqq \delta(1) + e + 1.56\gamma$. Therefore $\Omega \subset D(\delta(1) + (e+1.56\gamma))$. This completes the proof.

<u>Remark</u>. If $d\nu = \chi_W dm + d\mu$ for some measurable set W contained in $\bar{D}$ and for some finite positive measure μ, then, replacing $\hat{\nu}$ and $\hat{\chi}_\Omega$ by $\hat{\mu}$ and $\hat{\chi}_{\Omega\setminus W}$, we obtain $\Omega \subset D(\delta(\sqrt{\|\mu\|}) + (1/(2e)+e+1.56\gamma)\sqrt{\|\mu\|})$ for every $\Omega \in Q(\nu, AL^1)$.

<u>Corollary 6.5</u>. If $\text{supp}\ \nu \subset \overline{\Delta(\rho;c)}$, then $\Omega \subset \Delta(\rho + 5.3\sqrt{\|\nu\|};c)$ for every $\Omega \in Q(\nu, AL^1)$.

Proof. Let us consider a one-parameter family of domains $D(r) = D(r;\Delta(\rho;c),(z-c)/\rho)$. Then, by Lemma 6.3, $\gamma = 1$ and $\delta(\sqrt{\|\nu\|}) = \sqrt{\|\nu\|}$. It is easy to show that $D(r) = \Delta(\rho+r;c)$ for every $r \geqq 0$. Since $1 + e + 1.56 < 5.3$, we have the corollary.

Needless to say, the value 5.3 is not the smallest one. It is plausible that the best is equal to $1/\sqrt{\pi}$. For every $\varepsilon > 0$, we can find $N(\varepsilon)$ such that if $\sqrt{\|\nu\|} \geqq N(\varepsilon)\rho$, then $\Omega \subset \Delta(\rho + (1/\sqrt{\pi}+\varepsilon)\sqrt{\|\nu\|};c)$ for every $\Omega \in Q(\nu,AL^1)$ (see Proposition 10.23).

Let $R(\rho;c) = \{z \in \mathbb{C} \mid 1-\rho < |z-c| < 1+\rho\}$ for $\rho > 0$ and set $D(r) = D(r;R(\rho;c),\ (z-c)/(1+\rho),\ (z-c)/(1-\rho))$. Then $D(r) = R(\rho+r;c)$ if $0 \leqq r \leqq 1 - \rho$ and $D(r) = \Delta(1+\rho+r;c)$ if $r > 1 - \rho$. Since $1/(2e) < 0.19$, we have

Corollary 6.6. Suppose $\operatorname{supp} \nu \subset \overline{R(\rho;c)}$ and $\Omega \in Q(\nu,AL^1)$. Then $\Omega \subset R(\rho+5.3\sqrt{\|\nu\|};c)$ if $\sqrt{\|\nu\|} \leqq (1-\rho)/5.5$ and $\Omega \subset \Delta(1+\rho+5.3\sqrt{\|\nu\|};c)$ if $\sqrt{\|\nu\|} > (1-\rho)/5.5$.

Now let us deal with the second property of quadrature domains.

Theorem 6.7. Let $\Omega \in Q(\nu,AL^1)$ and $\operatorname{supp} \nu \subset \Omega$. Then every isolated nondegenerate boundary component is an analytic quasi-simple curve.

Proof. Let γ be such a boundary component, let D be the connected component of γ^c containing Ω and let $w = \varphi(z)$ be a one-to-one conformal mapping from D onto the unit disc Δ. Let us denote by w^* the reflection point of w with respect to $\partial\Delta$, set $V = \varphi(\Omega\setminus\text{supp } \nu)$ and $V^* = \{w^* \mid w \in V\}$.

Since γ is isolated, V is a boundary neighborhood of Δ. Let us define an analytic function f on $V \cup V^*$ by

$$f(w) = \begin{cases} \varphi^{-1}(w) & \text{on} \quad V \\ \overline{\{\frac{1}{\pi}(\hat{\chi}_\Omega + \pi\overline{z}) - \frac{1}{\pi}\hat{\nu}\}\circ\varphi^{-1}(w^*)} & \text{on} \quad V^*. \end{cases}$$

By Theorem 6.4, f is bounded near $\partial\Delta$ and so $\lim_{r\uparrow 1} f(re^{i\theta})$ and $\lim_{r\downarrow 1} f(re^{i\theta})$ exist almost everywhere on $\partial\Delta$. Let $w \in V$ and set $z = \varphi^{-1}(w)$. Then

$$f(w) - f(w^*) = \overline{\frac{1}{\pi}\left(\hat{\chi}_\Omega(z) - \hat{\nu}(z)\right)}.$$

Hence $\lim_{r\uparrow 1} f(re^{i\theta}) = \lim_{r\downarrow 1} f(re^{i\theta})$ a.e. on $\partial\Delta$.

From the generalized Painlevé theorem, it follows that f can be extended analytically onto $V \cup \partial\Delta \cup V^*$. Since f is univalent on V, by definition, $\gamma = f(\partial\Delta)$ is an analytic quasi-simple curve.

Let z_j, $j = 1,\cdots,m$, be points in $\mathbb{C}$ and a_{jk}, $j = 1,\cdots,m$, $k = 0, 1,\cdots,n_j$ be complex numbers. For every analytic function f defined in a fixed neighborhood of $\{z_1,\cdots,z_m\}$, we consider the linear functional

$$\sum_{j=1}^{m} \sum_{k=0}^{n_j} a_{jk}\, f^{(k)}(z_j).$$

Then, by using the Cauchy integral formula, we can find its complex representing measure whose support is contained an arbtrarily small neighborhood of $\{z_1,\cdots,z_m\}$. For the boundary $\partial\Omega$ of a quadrature domain Ω of this measure, the following remarkable result was obtained by Aharonov-Shapiro [1], Davis [9] and Gustafsson [11]: $\partial\Omega$ is contained an algebraic set; more precisely, for each quadrature domain Ω of the above measure, there exists a polynomial $p(x,y)$ with real coefficients and irreducible over the complex field such that $\partial\Omega \subset \{z=x+iy\in\mathbb{C} | p(x,y)=0\}$.

Finally we discuss the third property of quadrature domains. From the definition of $[\Omega]$, we only see $m([\Omega]\setminus\Omega) = 0$. For quadrature domains Ω, we have the following theorem:

<u>Theorem 6.8</u>. Let ν be a finite positive measure. If $\Omega \in Q(\nu,AL^1)$ and supp $\nu \subset \Omega$, then $[\Omega]\setminus\Omega$ is contained in a real analytic set in $[\Omega]\setminus$supp ν.

<u>Proof</u>. Let $E = \{z \in [\Omega]\setminus\text{supp }\nu |\ (\hat{\nu}-\hat{\chi}_{[\Omega]})(z) = 0\}$. Since $m([\Omega]\setminus\Omega) = 0$, $\hat{\chi}_{[\Omega]} = \hat{\chi}_\Omega$ so that $[\Omega]\setminus\Omega \subset E$. Hence the theorem follows, because $\hat{\nu} - \hat{\chi}_{[\Omega]}$ is real analytic and satisfies $\partial(\hat{\nu}-\hat{\chi}_{[\Omega]})/\partial\bar{z} = \pi \neq 0$ on $[\Omega]\setminus$supp ν. The dimension of E is at most one.

<u>§7. Existence of minimal quadrature domains</u>

Let Ω be a domain (or an open set) in $Q(\nu,F)$. We call Ω a minimal quadrature domain (or open set) in $Q(\nu,F)$ if $\Omega \subset G$ for every $G \in Q(\nu,F)$ satisfying supp $\nu \subset G$ and $[G] = [\Omega]$. In

Theorem 3.5, we have constructed the minimum domain $\tilde{W}$ in $Q(\nu,SL^1)$ for a finite positive measure ν on the closure of a bounded domain W with quasi-smooth boundary such that $d\nu/dm \geqq \chi_W$ and $\int d\nu > m(W)$. The minimum domain is of course minimal in $Q(\nu,SL^1)$.

In this section we shall prove the existence of minimal quadrature domains in $Q(\nu,HL^1)$ or $Q(\nu,AL^1)$.

Let ν be a complex measure with compact support and let Ω be a domain in $Q(\nu,AL^1)$ such that $\text{supp}\,\nu \subset \Omega$. We define a set $E(\Omega;\nu,AL^1)$ by

$$E(\Omega;\nu,AL^1) = \{\zeta \in \Omega\setminus\text{supp}\,\nu \mid \hat{\nu}(\zeta) = \hat{\chi}_\Omega(\zeta)\}.$$

For a domain Ω in $Q(\nu,HL^1)$, we also define a set $E(\Omega;\nu,HL^1)$ by

$$E(\Omega;\nu,HL^1) = \left\{\zeta \in \Omega\setminus\text{supp}\,\nu \,\middle|\, \int \text{Re}\,\frac{1}{z-\zeta}\,d\nu(z) = \int_\Omega \text{Re}\,\frac{1}{z-\zeta}\,dm(z),\right.$$

$$\int \text{Im}\,\frac{1}{z-\zeta}\,d\nu(z) = \int_\Omega \text{Im}\,\frac{1}{z-\zeta}\,dm(z)$$

$$\left.\text{and} \int \log|z-\zeta|\,d\nu(z) = \int_\Omega \log|z-\zeta|\,dm(z)\right\}.$$

It is clear that $E(\Omega;\nu,HL^1) \subset E(\Omega;\nu,AL^1)$ for $\Omega \in Q(\nu,HL^1)$ and $E(\Omega;\nu,AL^1) \subset E([\Omega];\nu,AL^1)$.

We have already seen, in Theorem 6.8, that if $\Omega \in Q(\nu,AL^1)$ and $\text{supp}\,\nu \subset \Omega$, then $[\Omega]\setminus\Omega \subset E = E([\Omega];\nu,AL^1)$ and $E([\Omega];\nu,AL^1)$ is a real analytic set in $[\Omega]\setminus\text{supp}\,\nu$.

The following lemma is deduced from the result due to Bers [5]:

Lemma 7.1. The class of linear combinations of $1/(z-\zeta_j)$ for $\zeta_j \in \partial\Omega$ is dense in $AL^1(\Omega)$. In other words, the class of rational functions having only simple poles outside Ω is dense in $AL^1(\Omega)$.

Proposition 7.2. If $\Omega \in Q(\nu, AL^1)$, supp $\nu \subset \Omega$ and supp $\nu \cap \overline{E([\Omega];\nu,AL^1)} = \phi$, then $[\Omega]\backslash E([\Omega];\nu,AL^1)$ is a minimal quadrature domain (or open set) in $Q(\nu,AL^1)$.

To show the existence of minimal quadrature domains for class HL^1, we prepare the following lemma:

Lemma 7.3. Let Ω be a bounded domain. Then the class of linear combinations of $\mathrm{Re}(1/(z-\zeta_j))$, $\mathrm{Im}(1/(z-\zeta_j))$ and $\log|z-\zeta_j|$ for $\zeta_j \in \partial\Omega$ is dense in $HL^1(\Omega)$.

Proof. To prove the lemma, it is enough to show that if a real valued function $g \in L^\infty(\Omega)$ satisfies $\int_\Omega \frac{g(z)}{z-\zeta}\,dm(z) = \int_\Omega \left(\log \frac{1}{|z-\zeta|}\right) g(z)\,dm(z) = 0$ for every $\zeta \in \partial\Omega$, then $\int_\Omega hg\,dm = 0$ for every $h \in HL^1(\Omega)$.

Set

$$U^g(\zeta) = \int_\Omega \left(\log \frac{1}{|z-\zeta|}\right) g(z)\,dm(z)$$

and let $\{\omega_n\}_{n=1}^\infty$ be a sequence of C^∞-functions on Ω such that $0 \leq \omega_n \leq 1$, $\omega_n = 0$ in a neighborhood of $\partial\Omega$, $\omega_n = 1$ outside a neighborhood of $\partial\Omega$, $\lim_{n\to\infty} \omega_n(z) = 1$ for all $z = x_1 + ix_2 \in \Omega$, and

$$|D^\alpha \omega_n(z)| \leq A_\alpha n^{-1} \delta(z)^{-|\alpha|} \left(\log \frac{1}{\delta(z)}\right)^{-1}$$

for all $z \in \Omega$ and all multi-indices α, where $D^{\alpha}\omega_n$ denotes the partial derivatives of ω_n, A_α denotes a constant depending only on α, $\alpha = (\alpha_1, \alpha_2)$, $|\alpha| = \alpha_1 + \alpha_2$ and $\delta(z)$ denote the minimum of e^{-2} and the distance from z to $\partial\Omega$. For the existence of the above sequence $\{\omega_n\}$, see Hedberg [12, p.13, Lemma 4].

Since $\partial U^g/\partial z = \hat{g}/2$ and $\partial\hat{g}/\partial\bar{z} = -\pi g$ in the sense of distributions, we have

$$\frac{\partial^2}{\partial\bar{z}\partial z}(\omega_n U^g) = \frac{\partial^2\omega_n}{\partial\bar{z}\partial z} U^g + \frac{1}{2}\frac{\partial\omega_n}{\partial z}\bar{\hat{g}} + \frac{1}{2}\frac{\partial\omega_n}{\partial\bar{z}}\hat{g} - \frac{\pi}{2}\omega_n g.$$

By the assumption, $\hat{g} = U^g = 0$ on $\partial\Omega$ so that

$$|\hat{g}(z)| = O\left(\delta(z)\log\frac{1}{\delta(z)}\right),$$

$$|U^g(z)| = O\left(\delta^2(z)\log\frac{1}{\delta(z)}\right)$$

in a neighborhood of $\partial\Omega$. Hence, by the above estimation of the derivatives of ω_n, we have

$$\int_\Omega hg\,dm = -\frac{2}{\pi}\lim_{n\to\infty}\int_\Omega h\frac{\partial^2}{\partial\bar{z}\partial z}(\omega_n U^g)dm$$

$$= -\frac{2}{\pi}\lim_{n\to\infty}\int_\Omega\left(\frac{\partial^2}{\partial z\partial\bar{z}}h\right)\omega_n U^g dm$$

$$= 0.$$

This completes the proof.

By Lemma 7.3, we have

<u>Proposition 7.4</u>. If $\Omega \in Q(\nu, HL^1)$, supp $\nu \subset \Omega$ and supp $\nu \cap \overline{E([\Omega];\nu,HL^1)} = \phi$, then $[\Omega]\backslash E([\Omega];\nu,HL^1)$ is a minimal quadrature domain (or open set) in $\Omega(\nu,HL^1)$.

From Example 1.2, we see that the minimum domain $\tilde{W}$ in $Q(\nu,SL^1)$ is not necessarily minimal in $Q(\nu,AL^1)$ and a minimal domain in $Q(\nu,HL^1)$ is not necessarily minimal in $Q(\nu,AL^1)$. In contrast to these facts, we have the following theorem:

Theorem 7.5. Let ν be a finite positive measure as in Theorem 3.4. Then the minimum domain $\tilde{W}$ in $Q(\nu,SL^1)$ is minimal in $Q(\nu,HL^1)$.

Proof. Let $S(G)$ be the class of all functions $s \in SL^1(G)$ each of which is bounded from above except a neighborhood of a some boundary point p of G and satisfies $s(z) = O(\log(1/|z-p|))$ on the neighborhood. Then, by the argument given in the proof of Theorem 3.5, it follows that $\tilde{W} \subset G$ for every $G \in Q(\nu,S)$. Hence $Q(\nu,S) = Q(\nu,SL^1)$ and $\tilde{W}$ is the minimum domain in $Q(\nu,S)$.

Assume that $\tilde{W}\setminus\{p\} \in Q(\nu,HL^1)$ for some $p \in \tilde{W}$. If $s \in S(\tilde{W}\setminus\{p\})$ is bounded from above in a neighborhood of p, then s can be extended to a function in $S(\tilde{W})$ so that $\int s d\nu \leqq \int_{\tilde{W}\setminus\{p\}} s dm$. If $s \in S(\tilde{W}\setminus\{p\})$ is unbounded, take $c > 0$ so that $s(z) \leqq c \log(1/|z-p|)$ in a neighborhood of p. Then $c \log(1/|z-p|) \in HL^1(\tilde{W}\setminus\{p\})$ and $s(z) - c \log(1/|z-p|) \in S(\tilde{W}\setminus\{p\})$, because $\tilde{W}$ is bounded by Theorem 6.4. Hence we have again $\int s d\nu \leqq \int_{\tilde{W}\setminus\{p\}} s dm$ so that $\tilde{W}\setminus\{p\} \in Q(\nu,S)$. This is a contradiction. Hence $\tilde{W}$ is minimal in $Q(\nu,HL^1)$.

§8. Relations between quadrature domains for classes SL^1, HL^1 and AL^1.

It always holds that quadrature domains for class SL^1 is one for class HL^1 and domains for class HL^1 is for class AL^1. Under what conditions the converse is true? We treat this problem in this section.

For the sake of simplicity, we consider a bounded domain Ω and let Ω^* be the Kerékjártó-Stoïlow compactification of Ω. Let π be the continuous mapping from $\mathbb{C}$ onto Ω^* such that $\pi|\Omega$ is the identity mapping on Ω. If two points p and q are in Ω^c, then $\pi(p) = \pi(q)$ if and only if p and q are in the same connected component of Ω^c.

Let us take $\{\zeta_j\}_{j=1}^{\infty} \subset \Omega^c$ so that $\{\pi(\zeta_j)\}$ is dense in $\Omega^*\backslash\Omega = \pi(\Omega^c)$ and $\{\zeta_j\}$ is dense in every connected component Γ of Ω^c such that the one-dimensional Hausdorff measure $\Lambda_1(\gamma)$ of some connected component γ of $\Gamma\backslash\overline{\Gamma^\circ}$ is infinite.

For a finite real measure ν with compact support, we set $U^\nu(\zeta) = \int\log(1/|z-\zeta|)d\nu(z)$. By using these notation, we have

<u>Proposition 8.1</u>. Let $\Omega \in Q(\nu,AL^1)$ and supp $\nu \subset \Omega$. Let $\{\zeta_j\}_{j=1}^{\infty} \subset \Omega^c$ be a sequence of points satisfying the above condition. Then $\Omega \in Q(\nu,HL^1)$ if and only if $U^\nu(\zeta_j) = U^{\chi_\Omega}(\zeta_j)$ for every j.

<u>Proof</u>. Let $u(z) = U^{\chi_\Omega}(z) - U^\nu(z)$. To prove the proposition, it is enough to show that if $u(\zeta_j) = 0$ for every j, then $\Omega \in Q(\nu,HL^1)$.

Let Γ be a connected component of Ω^c such that $\Lambda_1(\gamma) < +\infty$ for every connected component γ of $\Gamma\setminus\overline{\Gamma^\circ}$. We shall show that u is constant on Γ. Since u is harmonic on Γ°, is continuous on Γ and satisfies $\partial u/\partial z = (\hat{X}_\Omega - \hat{\nu})/2 = 0$ on Γ°, u is constant on every connected component of $\overline{\Gamma^\circ}$. Next take a connected component γ of $\Gamma\setminus\overline{\Gamma^\circ}$. By definition, for enery $\varepsilon > 0$, there is a sequence $\{\Delta(r_k;c_k)\}$ such that $\gamma \subset \cup\Delta(r_k;c_k)$, $r_k \leqq \varepsilon$ and $\Sigma r_k \leqq \Lambda_1(\gamma) + \varepsilon$. Let p and q be points in γ. Since

$$|\hat{X}_\Omega - \hat{\nu}| \leqq C\ \varepsilon \log \frac{1}{\varepsilon}$$

on $\cup\Delta(r_k;c_k)$ for sufficiently small ε,

$$\begin{aligned} |u(p)-u(q)| &\leqq 2(\Sigma r_k) C\ \varepsilon \log \frac{1}{\varepsilon} \\ &\leqq 2\bigl(\Lambda_1(\gamma)+\varepsilon\bigr) C\ \varepsilon \log \frac{1}{\varepsilon}. \end{aligned}$$

Hence $u(p) - u(q) = 0$ so that u is constant on $\overline{\gamma}$ for every connected component γ of $\Gamma\setminus\overline{\Gamma^\circ}$. Since $\overline{\gamma} \cap \overline{\Gamma^\circ} \neq \phi$ if $\gamma \neq \phi$ and $\Gamma^\circ \neq \phi$, and since Γ° has at most a countable number of connected components, u is constant on Γ.

Therefore if $u(\zeta_j) = 0$ for every j, then $u = 0$ on $\partial\Omega$. By Lemma 7.3, it follows that $\Omega \in Q(\nu, HL^1)$.

From Proposition 8.1, the following corollary immediately follows, because each nondegenerate boundary component is an analytic quasi-simple curve by Theorem 6.7.

<u>Corollary 8.2</u>. Let ν be a finite real measure with compact support and let Ω be a finitely connected domain in $Q(\nu,AL^1)$ with supp $\nu \subset \Omega$. Then $\Omega \in Q(\nu,HL^1)$ if and only if there exists a point ζ on every bounded connected component of Ω^c such that $U^\nu(\zeta) = U^{\chi_\Omega}(\zeta)$. In particular, if Ω is a simply connected domain in $Q(\nu,AL^1)$ with supp $\nu \subset \Omega$, then $\Omega \in Q(\nu,HL^1)$.

<u>Proposition 8.3</u>. Let ν be a finite positive measure with compact support and let Ω be a domain in $Q(\nu,HL^1)$ with supp $\nu \subset \Omega$. Then $\Omega \in Q(\nu,SL^1)$ if and only if $U^\nu(z) \geqq U^{\chi_\Omega}(z)$ on Ω.

<u>Proof</u>. Since $Q(\nu,SL^1) = Q(\nu,S)$ (for the definition of class S, see the proof of Theorem 7.5), it is enough to show that

$$\int s d\nu \leqq \int_\Omega s dm$$

for every $s \in S(\Omega)$ if $U^\nu(z) \geqq U^{\chi_\Omega}(z)$ on Ω.

Let $s \in S(\Omega)$ be bounded from above except a neighborhood of some $p \in \partial\Omega$ and let $s(z) \leqq c \log(1/|z-p|)$ on the neighborhood. Then $v(z) = s(z) - c \log(1/|z-p|)$ is bounded from above and is of class SL^1. Hence there is a least harmonic majorant h of v and so we have the Riesz decomposition of v:

$$v(z) = h(z) - \int_\Omega g(\zeta;z,\Omega)d\mu(\zeta),$$

where μ is a positive measure on Ω and g is the Green function on Ω with pole at z. Since v and h are of class $L^1(\Omega)$, we have $-\int_\Omega g(\zeta;z,\Omega)d\mu(\zeta) \in SL^1(\Omega)$. If $U^\nu(z) \geqq U^{\chi_\Omega}(z)$ on Ω, then

$$\int g(\zeta;z,\Omega)d\nu(\zeta) \geqq \int g(\zeta;z,\Omega)dm(\zeta)$$

on Ω so that

$$\int s d\nu \leqq \int_\Omega s dm$$

for every $s \in S(\Omega)$.

<u>Corollary 8.4</u>. Let ν be a finite positive measure with compact support and let Ω be a domain in $Q(\nu,HL^1)$ with supp $\nu \subset \Omega$ and $U^\nu(z) \geqq U^{\chi_\Omega}(z)$ in a neighborhood of supp ν.

If a minimal domain G in $Q(\nu,HL^1)$ with $[G] = [\Omega]$ is also minimal in $Q(\nu,AL^1)$, namely, $\hat{\nu}(z) \neq \hat{\chi}_\Omega(z)$ on $G \backslash$supp ν, then $\Omega \in Q(\nu,SL^1)$.

<u>Proof</u>. Assume $G \notin Q(\nu,SL^1)$. Then, by Proposition 8.3, $u(z) = U^{\chi_\Omega}(z) - U^\nu(z) = U^{\chi_G}(z) - U^\nu(z) > 0$ for a point $z \in G \backslash$ supp ν. Since u is continuous on $\overline{G}\backslash$supp ν, $u(z) \leqq 0$ in a neighborhood of supp ν and $u(z) \to 0$ as $z \to \partial G$, u attains its maximum at a point $\zeta \in G\backslash$supp ν. Hence $\hat{\chi}_\Omega(\zeta) - \hat{\nu}(\zeta) = 2(\partial u/\partial z)(\zeta) = 0$ so that G is not minimal in $Q(\nu,AL^1)$.

§9. <u>Uniqueness in the strict sense</u>

In general, there are many domains in $Q(\nu,AL^1)$. Under what conditions on ν is the domain in $Q(\nu,AL^1)$ determined uniquely? In this section we deal with the problem and give sufficient conditions. First we show

<u>Proposition 9.1</u>. Let ν be a measure as in Theorem 3.4. Let Ω be a domain in $Q(\nu, HL^1)$ containing supp ν. If the maximum domain $[\tilde{W}]$ in $Q(\nu, SL^1)$ is a Carathéodory domain, namely, if $\partial[\tilde{W}]$ coincides with the boundary of the unbounded component of $[\tilde{W}]^e$, then $\Omega \subset [\tilde{W}]$ and $\Omega \in Q(\nu, SL^1)$.

<u>Proof</u>. By Corollary 4.10, $\bar{\Omega} \subset (\tilde{W}^e)^c = ([\tilde{W}]^e)^c = \overline{[\tilde{W}]}$. Since, by Theorem 6.7, $\partial[\tilde{W}]$ is an analytic quasi-simple curve, $\Omega \subset (\bar{\Omega})^\circ \subset (\overline{[\tilde{W}]})^\circ = [\tilde{W}]$. The proposition follows from Theorem 7.5.

The following corollary follows immediately from Corollary 8.2:

<u>Corollary 9.2</u>. If the maximum domain $[\tilde{W}]$ in $Q(\nu, SL^1)$ is a Carathéodory domain, then it is the unique simply connected domain in $Q(\nu, AL^1)$ containing supp ν.

<u>Proposition 9.3</u>. Let ν be a measure as in Theorem 3.4. Let Ω be a simply connected domain in $Q(\nu, AL^1)$ containing supp ν and satisfying $\hat{\nu} \neq \hat{\chi}_\Omega$ on $\Omega \backslash$supp ν. Then Ω is the unique domain in $Q(\nu, SL^1)$.

<u>Proof</u>. By Corollary 8.2, $\Omega \in Q(\nu, HL^1)$. Since $\hat{\nu} \neq \hat{\chi}_\Omega$ on $\Omega \backslash$supp ν, Ω is minimal in $Q(\nu, AL^1)$. Hence, by Corollary 8.4, $\Omega \in Q(\nu, SL^1)$.

If we add further conditions on Ω in Proposition 9.3, we obtain an interesting result (see Corollary 9.5). More generally we have

Proposition 9.4. Let ν be a complex measure with compact support. Let Ω be a finitely connected domain in $Q(\nu,AL^1)$ containing supp ν such that $\partial\Omega = \partial(\Omega^e)$ and each connected component Ω_j^e, $j = 1,\cdots,n$, of Ω^e is a domain with quasi-smooth boundary. Let G be a domain in $Q(\nu,AL^1)$ containing supp ν and satisfying $\Omega_j^e \cap G \neq \phi$ for every j, $j = 1,\cdots,n$. Then $E(\Omega;\nu,AL^1)\setminus G \neq \phi$ (for the definition, see §7).

Proof. By Theorem 6.4, Ω and G are both bounded. Take $\Delta = \Delta(r;0)$ so that $\overline{\Omega} \cup \overline{G} \subset \Delta$. By the assumption on G, $W = (\Delta\setminus\overline{\Omega}) \cup G$ is a bounded domain. Let $\tilde{W}$ be the minimum domain in $Q(\chi_{\Delta\setminus\overline{\Omega}}m+\chi_G m,SL^1)$. Since $m((\Delta\setminus\overline{\Omega}) \cap G) = m(\Omega^e \cap G) > 0$ and $\partial\Omega$ consists of a finite number of analytic simple curves (see Theorem 6.7), by an argument similar to Theorem 3.4, we see that $(\partial\Delta) \cup (\partial\Omega) \subset \tilde{W}$. Let μ be the positive measure on $\partial\Delta$ defined by $d\mu = (1/(2\pi))\cdot d\arg z$ and let $\tilde{W}(t)$ be the minimum domain in $Q(\chi_{\Delta\setminus\overline{\Omega}}m + \chi_G m + t\mu, SL^1)$ for $t \geqq 0$. Since $m(\Delta\setminus\tilde{W}(0)) = m(\Delta\setminus\tilde{W}) > 0$, $\Delta\setminus\tilde{W}(t) = \phi$ for large t and $0 \leqq m(\Delta\setminus\tilde{W}(s)) - m(\Delta\setminus\tilde{W}(t)) \leqq m(\tilde{W}(t)\setminus\tilde{W}(s)) = t - s$ for each pair of numbers s and t with $0 \leqq s \leqq t$, there is $t_0 > 0$ such that $m(\Delta\setminus\tilde{W}(t)) > 0$ for $t < t_0$ and $m(\Delta\setminus\tilde{W}(t)) = 0$ for $t \geqq t_0$. Since $\Delta\setminus\tilde{W}(t)$ is closed and decreases as t increases, and $\Delta\setminus\tilde{W}(t_0) = \cap_{t<t_0} \Delta\setminus\tilde{W}(t)$ by Corollary 3.9, $\Delta\setminus\tilde{W}(t_0)$ is not empty.

Next we shall show that $\Delta \cup \tilde{W}(t_0) = \Delta(\sqrt{r^2+t_0/\pi};0)$. Let $\zeta \notin \Delta \cup \tilde{W}(t_0)$. Since $m(\partial\Omega) = 0$,

$$\hat{\chi}_{\Delta\cup\tilde{W}(t_0)}(\zeta) = \hat{\chi}_{\tilde{W}(t_0)}(\zeta)$$

$$= \hat{\chi}_{\Delta\setminus\overline{\Omega}}(\zeta) + \hat{\chi}_G(\zeta) + t_0\hat{\mu}(\zeta)$$

$$= \hat{\chi}_{\Delta}(\zeta) - t_0/\zeta$$

$$= -(\pi r^2 + t_0)/\zeta.$$

Hence, by the argument given in Example 1.1, $\Delta \cup \tilde{W}(t_0) = \Delta(\sqrt{r^2+t_0/\pi};0)$.

This implies $\hat{\chi}_{\tilde{W}(t_0)}(\zeta) = \hat{\chi}_{\Delta\cup\tilde{W}(t_0)}(\zeta) = -\pi\bar{\zeta}$ on Δ. Since $\hat{\mu}(\zeta) = 0$ on Δ, $\hat{\chi}_{\Omega}(\zeta) = \hat{\chi}_{\overline{\Omega}}(\zeta) = \hat{\chi}_G(\zeta) = \hat{\nu}(\zeta)$ on $\Delta\setminus\tilde{W}(t_0)$. Since $\Delta \setminus W \subset \overline{\Omega}$ and $\partial\Omega \subset \tilde{W} \subset \tilde{W}(t_0)$, $\Delta\setminus\tilde{W}(t_0) \subset \Omega\setminus\text{supp}\ \nu$. Hence $E(\Omega;\nu, AL^1) \supset \Delta\setminus\tilde{W}(t_0)$ so that $E(\Omega;\nu,AL^1)\setminus G \supset \Delta\setminus\tilde{W}(t_0) \neq \phi$.

<u>Corollary 9.5</u>. Let ν be a complex measure with compact support. Let Ω be a domain in $Q(\nu,AL^1)$ containing supp ν. If $\partial\Omega$ is a (piecewise) smooth simple curve such that angle $V_1 < 2\Pi$ on $\partial\Omega$ and if $\hat{\nu} \neq \hat{\chi}_{\Omega}$ on $\Omega\setminus\text{supp}\ \nu$, then Ω is the unique domain in $Q(\nu,AL^1)$ containing supp ν.

<u>Proof</u>. Let G be a domain in $Q(\nu,AL^1)$ containing supp ν. If $G \neq \Omega$, then $\Omega^e \cap G \neq \phi$. Hence, by Proposition 9.4, $E(\Omega;\nu, AL^1) \supset E(\Omega,\nu,AL^1)\setminus G \neq \phi$. This is a contradiction.

There are many applications of Corollary 9.5. Let δ_0 be the Dirac measure at 0 and $\Omega \in Q(\pi\delta_0,AL^1)$. In Example 1.1, we have shown that $\Delta(1;0)$ is the unique domain in $Q(\pi\delta_0,AL^1)$. This also follows from Corollary 9.5. By the mean-value property of harmonic functions, $\Delta(1;0) \in Q(\nu, AL^1)$. Since $\hat{\nu}(z) = -\pi/z$ on $\Delta(1;0)\setminus\{0\}$ and $\hat{\chi}_{\Delta(1;0)}(z) = -\pi\bar{z}$ on $\Delta(1;0)$, $\hat{\nu}(z) \neq \hat{\chi}_{\Delta(1;0)}(z)$ on $\Delta(1;0)\setminus\{0\}$. If $\Omega \in Q(\pi\delta_0, AL^1)$, then, by definition, $0 \in \Omega$. Hence, by Corollary 9.5, $\Delta(1;0)$ is the unique domain in $Q(\pi\delta_0, AL^1)$.

Next we shall show another example. The example was given by Davis [7] in 1969. To construct the quadrature domain, he made use of the Bergman reproducing kernel function. He assumed that the domain is symmetric with respect to the real axis and is simply connected. In Example 9.6 below, we shall show the domain is unique. Hence the assumption is unnecessary. For details, see §14.

Example 9.6 (Davis [7]). Let $\alpha = \sqrt{(e^{\pi}-1)/(e^{\pi}+1)} = 0.95768\cdots$ and let Ω be the image of $\{w \in \mathbb{C} \mid |w| < 1\}$ under $z = (1/\pi)\log((1+\alpha w)/(1-\alpha w))$, where $-\pi < \operatorname{Im}\log((1+\alpha w)/(1-\alpha w)) < \pi$. Then $w = w(z) = (1/\alpha)(e^{\pi z}-1)/(e^{\pi z}+1)$ and $\log((w+\alpha)/(w-\alpha))$ has a single-valued branch on $\mathbb{C}\setminus[-\alpha,\alpha]$. Therefore

$$\begin{aligned}
\int_{\Omega} f\,dm &= \frac{1}{2i}\int_{\partial\Omega} f(z)\bar{z}\,dz \\
&= \frac{1}{2i}\int_{\partial\Omega} f(z)\left\{\frac{1}{\pi}\log\frac{w(z)+\alpha}{w(z)-\alpha}\right\}dz \\
&= \frac{1}{2\pi i}\int_{-1}^{1} f(x)\left\{\log\left|\frac{w(x)+\alpha}{w(x)-\alpha}\right|+\pi i\right\}dx \\
&\qquad - \frac{1}{2\pi i}\int_{-1}^{1} f(x)\left\{\log\left|\frac{w(x)+\alpha}{w(x)-\alpha}\right|-\pi i\right\}dx \\
&= \int_{-1}^{1} f(x)\,dx
\end{aligned}$$

for every function f analytic on $\overline{\Omega}$. Hence, by Lemma 7.1, $\Omega \in Q(\nu, AL^1)$, where $d\nu = \chi_{[-1,1]}dx$.

Next let us show $\hat{\nu}(z) \neq \hat{\chi}_\Omega(z)$ on $\Omega\backslash[-1,1]$. It is easy to show that

$$\hat{\nu}(z) = \int_{-1}^{1} \frac{d\xi}{\xi - z} = \log\left|\frac{z-1}{z+1}\right| + i \arg\left(\frac{z-1}{z+1}\right)$$

on $\Omega\backslash[-1,1]$, where $-\pi < \arg((z-1)/(z+1)) < \pi$. Since $\hat{\chi}_\Omega(z)$ is harmonic on Ω, $\mathrm{Im}\,\hat{\chi}_\Omega = 0$ on $\Omega \cap \{z \in \mathbb{C}|\ \mathrm{Im}\, z = 0\}$ and $\mathrm{Im}\,\hat{\nu} = \mathrm{Im}\,\hat{\chi}_\Omega$ on $\partial\Omega$, we obtain $\mathrm{Im}\,\hat{\nu} \lessgtr \mathrm{Im}\,\hat{\chi}_\Omega$ on $\Omega \cap \{z \in \mathbb{C}|\ \mathrm{Im}\, z \lessgtr 0\}$. Since $\mathrm{Re}\,\hat{\nu} = \mathrm{Re}\,\hat{\chi}_\Omega = 0$ on $\Omega \cap \{z \in \mathbb{C}|\ \mathrm{Re}\, z = 0\}$, $\mathrm{Re}\,\hat{\nu} = \mathrm{Re}\,\hat{\chi}_\Omega$ on $\partial\Omega$, we obtain $\mathrm{Re}\,\hat{\nu} \lessgtr \mathrm{Re}\,\hat{\chi}_\Omega$ on $(\Omega\backslash[-1,1]) \cap \{z \in \mathbb{C}|\ \mathrm{Re}\, z \gtrless 0\}$. Hence $\hat{\nu}(z) \neq \hat{\chi}_\Omega(z)$ on $\Omega\backslash[-1,1]$.

Since $1/(z-\zeta) \in AL^1(G)$ for every $\zeta \in G^c$, by definition, $\mathrm{supp}\,\nu = [-1,1] \subset G$ for every $G \in Q(\nu, AL^1)$. Hence, by Corollary 9.5, we see that Ω is the unique domain in $Q(\nu, AL^1)$.

§10. Monotone increasing families of quadrature domains

We shall deal with monotone increasing families of quadrature domains in this section and show that there exists, essentially, at most one family for class HL^1 (see Theorem 10.13 and Corollary 10.14).

Let $\{\Omega(t)\}_{t \geqq 0}$ be a family of domains with $\Omega(0) \neq \phi$. Let us call it monotone increasing if $\Omega(s) \subset \Omega(t)$ for every pair of s and t with $0 \leqq s \leqq t$.

Let $\{\nu(t)\}_{t\geqq 0}$ be a family of finite positive measures with $\nu(0) \neq 0$. We call it monotone increasing if $\nu(s)(E) \leqq \nu(t)(E)$ for every pair of s and t with $0 \leqq s \leqq t$ and for every Borel set E.

We call $\{\Omega(t)\}$ the family of quadrature domains of a monotone increasing family $\{\nu(t)\}$ for class F if

(1) $\{\Omega(t)\}$ is monotone increasing.

(2) $\Omega(t) \in Q(\nu(t),F)$ for every $t \geqq 0$.

We denote the class of all such families by $Q(\{\nu(t)\},F)$.

We use the following notation: $\{\Omega_1(t)\} \subset \{\Omega_2(t)\}$ if $\Omega_1(t) \subset \Omega_2(t)$ for every $t \geqq 0$; $[\{\Omega_1(t)\}] = [\{\Omega_2(t)\}]$ if $[\Omega_1(t)] = [\Omega_2(t)]$ for every $t \geqq 0$.

For an open set $\Omega \neq \phi$ and $d \geqq 0$, we define open sets $\Omega \oplus d$ and $\Omega \ominus d$ by

$$\Omega \oplus d = \{z \in \mathbb{C} |\ d(z,\Omega) < d\},$$
$$\Omega \ominus d = \{z \in \Omega |\ d(z,\Omega^c) > d\}$$

if $\Omega \neq \mathbb{C}$ and $d > 0$, $\Omega \oplus d = \Omega \ominus d = \Omega$ otherwise.

For a monotone increasing family $\{\Omega(t)\}$ of domains, we define for each $t \geqq 0$ the distance function $d(h;t)$ on $[-t,\infty)$ by

$$d(h;t) = \begin{cases} -\inf\{d \geqq 0 |\ \Omega(t) \ominus d \subset \Omega(t+h)\}, & -t \leqq h < 0 \\ 0, & h = 0 \\ \inf\{d \geqq 0 |\ \Omega(t+h) \subset \Omega(t) \oplus d\}, & h > 0. \end{cases}$$

For each fixed $t \geqq 0$, $d(h;t)$ is monotone increasing. Let $d(\pm 0;t) = \lim_{h\to\pm 0} d(h;t)$, $L = \{t > 0 |\ d(-0;t) < 0\}$ and R =

$\{t \geqq 0 \mid d(+0;t) > 0\}$. It is easy to show that L and R are both at most countably infinite. If $L = R = \phi$, then we say the family is continuous with respect to distance.

For a monotone increasing family $\{\Omega(t)\}$ of domains, we also define the measure function $m(h;t)$ on $[-t,\infty)$ by $m(h;t) = m(\Omega(t+h)) - m(\Omega(t))$. For each $t \geqq 0$, $m(h;t)$ is monotone increasing. Let $m(\pm 0;h) = \lim_{h\to\pm 0} m(h;t)$. The sets $\{t > 0 \mid m(-0;t) < 0\}$ and $\{t \geqq 0 \mid m(+0;t) > 0\}$ are both at most countably infinite. If $m(-0;t) = 0$ for every $t > 0$ and $m(+0;t) = 0$ for every $t \geqq 0$, we call $\{\Omega(t)\}$ the continuous family with respect to measure.

For a monotone increasing family $\{\Omega(t)\}$ of domains, we set

$$\mathrm{disc}\{\Omega(t)\} = [\bigcup_{t\geqq 0} \Omega(t)\backslash\Omega(0)]\backslash \bigcup_{t\geqq 0} \partial\Omega(t).$$

Let us first show

Lemma 10.1. If $\{\Omega(t)\}$ is continuous with respect to distance (resp. with respect to measure), then $\mathrm{disc}\{\Omega(t)\} = \phi$ (resp. $m(\mathrm{disc}\{\Omega(t)\}) = 0$).

Proof. We define a function $t(p)$ on $D = \mathrm{disc}\{\Omega(t)\}$ by

$$t(p) = \sup\{t \geqq 0 \mid p \in \Omega(t)^e\}.$$

Then $p \in \Omega(s)^e$ for s with $0 \leqq s < t(p)$ and $p \in \Omega(s)$ for s with $s > t(p)$. Set

$$D(t,L) = \{p \in D \mid t(p) = t \text{ and } p \in \Omega(t)\},$$

$$D(t,R) = \{p \in D \mid t(p) = t \text{ and } p \in \Omega(t)^e\}$$

for every $t \geqq 0$.

If $p \in D(t,L)$, then $t > 0$ and $d(p,\partial\Omega(t)) > 0$. Hence $p \in \Omega(t) \ominus \delta$ for every δ with $0 < \delta < d(p,\partial\Omega(t))$. Since $p \in \Omega(s)^e$ if $s < t$, $\Omega(t) \ominus \delta \not\subset \Omega(s)$ for every δ with $0 < \delta < d(p,\partial\Omega(t))$ and s with $0 \leqq s < t$. Therefore $d(h;t) \leqq -d(p,\partial\Omega(t))$ for $h < 0$ so that $t \in L$. Thus we have proved that if $D(t,L) \neq \phi$, then $t \in L$. Similarly, it follows that if $D(t,R) \neq \phi$, then $t \in R$.

Since

$$D = \bigcup_{t\in L} D(t,L) \cup \bigcup_{t\in R} D(t,R)$$

and

$$D(t,L) \subset \Omega(t) \cap \bigcap_{s<t} \Omega(s)^e \subset \lim_{s\uparrow t}[\Omega(t)\backslash\Omega(s)],$$

$$D(t,R) \subset \Omega(t)^e \cap \bigcap_{s>t} \Omega(s) \subset \lim_{s\downarrow t}[\Omega(s)\backslash\Omega(t)],$$

we have the lemma.

For a monotone increasing family $\{\Omega(t)\}$ of domains, we define stationary points and stagnant points as follows: $p \in \cup_{t\geqq 0} \Omega(t)\backslash\Omega(0)$ is called a stationary point of $\{\Omega(t)\}$ if $p \in \partial\Omega(t)$ for every t on an open interval; $p \in \cup_{t\geqq 0} \Omega(t)\backslash\Omega(0)$ is called a stagnant point of $\{\Omega(t)\}$ if there is $t(p) \geqq 0$ and $\varepsilon(p) > 0$ such that $m(\Delta(r;p)\backslash\Omega(t)) > 0$ for every $r > 0$ and $t < t(p)$, and $m(\Delta(\varepsilon(p);p)\backslash\Omega(t)) = 0$ for every $t \geqq t(p)$. We denote by $\mathrm{stat}\{\Omega(t)\}$ (resp. $\mathrm{stag}\{\Omega(t)\}$) the set of all stationary points (resp. stagnant points) of $\{\Omega(t)\}$. It is easy to show that $\mathrm{disc}\{\Omega(t)\} \cap \mathrm{stat}\{\Omega(t)\} = \phi$ and $\cup_{t\in L} D(t,L) \subset \mathrm{stag}\{\Omega(t)\}$.

Lemma 10.2. If $\{\Omega(t)\}$ is continuous with respect to measure, then there is at least one stagnant point on every compact isolated connected component of $\cup_{t \geqq 0} \Omega(t)\setminus\Omega(0)$.

Proof. Let K be a compact isolated connected component of $\cup_{t \geqq 0} \Omega(t)\setminus\Omega(0)$. Let $K(t) = K\setminus\Omega(t)$ and set $t_0 = \inf\{t \geqq 0 | m(K(t)) = 0\}$. Since $K(t)$ is monotone decreasing and $K(t) = \phi$ for some $t > 0$, t_0 exists and is finite. If $t_0 > 0$, then $m(K(s)) > 0$ for $s < t_0$ so that $K(s) \neq \phi$. Hence $L = \cap_{s<t_0} K(s) \neq \phi$. If $t_0 = 0$, then $K(t_0) = K \neq \phi$. We set $L = K$. Since $\{\Omega(t)\}$ is continuous with respect to measure, $m(K(t_0)) = 0$. Hence, for every $p \in L$, we can find a disc $\Delta(\varepsilon(p);p)$ such that $m(\Delta(\varepsilon(p);p)\setminus\Omega(t)) = 0$ for every $t \geqq t_0$. Since L is compact, for every $r > 0$, we can choose discs $\Delta(r;p_j)$, $j = 1,\cdots,n$ with $p_j \in L$ so that $L \subset \cup\Delta(r;p_j)$. Take $s_0 < t_0$ so that $K(s_0) \subset \cup\Delta(r;p_j)$. Then $\Sigma m(\Delta(r;p_j)\setminus\Omega(s)) \geqq m(\cup\Delta(r;p_j)\setminus\Omega(s)) \geqq m(K(s)) > 0$ for evey s with $s_0 \leqq s < t_0$. Therefore we can choose $p_j \in L$ so that $m(\Delta(r;p_j)\setminus\Omega(s)) > 0$ for every $s < t_0$. We denote by $p(r)$ one of such a $p_j \in L$. Choose r_k, $k = 1, 2,\cdots$, so that $r_k \to 0$ as $k \to \infty$ and let p be an accumulation point of $\{p(r_k)\}$. Then $m(\Delta(r;p)\setminus \Omega(s)) > 0$ for every $r > 0$ and every $s < t_0$. Hence $p \in L \subset K$ is a stagnant point of $\{\Omega(t)\}$.

Now let us consider monotone increasing families of quadrature domains. Let us call $\{\nu(t)\}$ the continuous family if $\|\nu(t+h)\| - \|\nu(t)\| \to 0$ as $h \to 0$ for every $t \geqq 0$.

<u>Lemma 10.3</u>. If $\pm 1 \in F(\Omega(t))$ for every $t \geqq 0$, then $\{\Omega(t)\} \in Q(\{\nu(t)\},F)$ is continuous with respect to measure if and only if $\{\nu(t)\}$ is continuous.

<u>Proof</u>. The lemma follows immediately from the equation

$$m(h;t) = \int d\nu(t+h) - \int d\nu(t)$$
$$= \|\nu(t+h)\| - \|\nu(t)\|.$$

<u>Proposition 10.4</u>. Let $\{\nu(t)\}$ be a continuous family of measures with compact supports such that $\operatorname{supp} \nu(t) = \operatorname{supp} \nu(0)$ for every $t \geqq 0$. Then each $\{\Omega(t)\} \in Q(\{\nu(t)\},AL^1)$ is right continuous with respect to distance, namely, $d(+0;t) = 0$ for every $t \geqq 0$.

<u>Proof</u>. Assume $d(+0;t) > 0$ for some $t \geqq 0$. By definition, $\Omega(t+h)\setminus(\Omega(t) \oplus d) \neq \phi$ for a fixed d with $0 < d < d(+0;t)$ and for every $h > 0$. Take h so small that $\|\nu(t+h)-\nu(t)\|^{1/2} < d/5.5$ and let $c \in \Omega(t+h)\setminus(\Omega(t) \oplus d)$. Take $R > 0$ so that $\Omega(t) \subset \Delta(R;c)$ and set $D(r) = \Delta(R+r;c)\setminus\overline{\Delta(d-r;c)}$ for r with $0 \leqq r < d$. Since $\Omega(t) \subset D(0)$, by Remark to Theorem 6.4, we have $\Omega(t+h) \subset D((1+1/(2e)+e+1.56)\|\nu(t+h)-\nu(t)\|^{1/2}) \subset D(5.5\|\nu(t+h)-\nu(t)\|^{1/2}) \subset \Delta(R+d;c)\setminus\{c\}$. Hence $c \notin \Omega(t+h)$. This is a contradiction.

Next let us deal with families of quadrature domains for class SL^1. For the sake of simplicity, we deal with the following monotone increasing families $\{\nu(t)\}$ of finite positive measures: $\nu(t)$ is a measure on the closure $\overline{W}$ of a bounded domain W with quasi-smooth boundary such that $d\nu(t)/dm \geqq \chi_W$ and $\int d\nu(t) > m(W)$

for every $t \geqq 0$. For such a measure $\nu(t)$, we have constructed the minimum domain $\tilde{W}(t)$ in $Q(\nu(t),SL^1)$, see Theorems 3.4 and 3.5. By Lemma 3.6, $\tilde{W}(s) \subset \tilde{W}(t)$ if $s \leqq t$. Hence we have

Proposition 10.5. The class $Q(\{\nu(t)\},SL^1)$ is not empty and $\{\tilde{W}(t)\}$ is the minimum family in $Q(\{\nu(t)\},SL^1)$. The family is unique in the sense that $[\{\Omega(t)\}] = [\{\tilde{W}(t)\}]$ for every family $\{\Omega(t)\}$ in $Q(\{\nu(t)\},SL^1)$.

Proposition 10.6. If $\{\nu(t)\}$ is continuous, then the minimum family $\{\tilde{W}(t)\}$ in $Q(\{\nu(t)\},SL^1)$ is continuous with respect to distance.

Proof. By Proposition 10.4, $\{\tilde{W}(t)\}$ is right continuous. By Corollary 3.9, $\cup_{s<t} \tilde{W}(s) = \tilde{W}(t)$ for every $t > 0$. Hence $\{\tilde{W}(t)\}$ is also left continuous.

Proposition 10.7. If $\{\nu(t)\}$ is strictly increasing, namely, if $\|\nu(s)\| < \|\nu(t)\|$ for every pair of numbers s and t with $0 \leqq s < t$, and if $\{\Omega(t)\} \in Q(\{\nu(t)\},SL^1)$, then, for every $\varepsilon > 0$, there is $\delta = \delta(\varepsilon,t) > 0$ such that $\Omega(t) \oplus \delta \subset \Omega(t+\varepsilon)$.

Proof. By using the same argument as in the beginning of the proof of Theorem 3.4, we may assume that $\mathrm{supp}(\nu(t)-\chi_W m) \subset W$ fo every $t \geqq 0$. Let $\nu(t) = \chi_W m + \mu$ and $\nu(t+\varepsilon) = \chi_W m + \mu + \xi$. Let us write $\tilde{W}(\nu)$ the minimum domain in $Q(\nu,SL^1)$ and set $W + p = \{w+p|\ w \in W\}$. Since $\overline{W} \subset \tilde{W}(\chi_W m+\xi/2)$, we can find $\delta > 0$ such that $W \oplus \delta \subset \tilde{W}(\chi_W m+\xi/2)$. Take p with $|p| < \delta$ and set $\varphi(z) = z + p$.

Then $\chi_W m\circ\varphi^{-1} = \chi_{W+p}m$ and $\tilde{W}(\chi_{W+p}m + \mu\circ\varphi^{-1}) = \tilde{W}(\chi_W m + \mu) + p = \tilde{W}(t) + p$. Therefore $\tilde{W}(t) + p \subset \tilde{W}(\chi_W m + \xi/2 + \mu\circ\varphi^{-1})$ for every p with $|p| < \delta$. Hence $\tilde{W}(t) \oplus \delta \subset \tilde{W}(\chi_W m + \xi/2 + \mu\circ\varphi^{-1})$. Since $\operatorname{supp}\mu \subset W$, there is $C(\delta) > 1$ such that $\beta(\mu\circ\varphi^{-1},W) \leqq C(\delta)\beta(\mu,W)$ for δ with $\operatorname{supp}(\mu\circ\varphi^{-1}) \subset W$. Since $\operatorname{supp}\xi \subset W$ and we can choose $C(\delta)$ so that $C(\delta) \to 1$ as $\delta \to 0$, we can find δ such that $\beta(\mu\circ\varphi^{-1}, W) \leqq \beta(\mu,W) + \beta(\xi/2,W)$. For such a δ, we have $\tilde{W}(t) \oplus \delta \subset \tilde{W}(\chi_W m + \xi/2 + \mu\circ\varphi^{-1}) \subset \tilde{W}(\chi_W m + \xi/2 + \mu + \xi/2) = \tilde{W}(t+\varepsilon)$. Hence $\Omega(t) \oplus \delta = \tilde{W}(t) \oplus \delta \subset \tilde{W}(t+\varepsilon) \subset \Omega(t+\varepsilon)$.

Corollary 10.8. If $\{\nu(t)\}$ is strictly increasing, then $\operatorname{stat}\{\Omega(t)\} = \phi$ for every $\{\Omega(t)\} \in Q(\{\nu(t)\},SL^1)$.

Corollary 10.9. Let $\{\nu(t)\}$ be strictly increasing and continuous, and let $\{\tilde{W}(t)\}$ be the minimum family in $Q(\{\nu(t)\}, SL^1)$. Then, for every $p \in \cup_{t\geqq 0} \tilde{W}(t)\setminus\tilde{W}(0)$, there is a uniquely determined $t = t(p) \geqq 0$ such that $p \in \partial\tilde{W}(t)$. The function t is continuous on $\cup_{t\geqq 0} \tilde{W}(t)\setminus\tilde{W}(0)$.

Proof. By Lemma 10.1 and Proposition 10.6, $\operatorname{disc}\{\tilde{W}(t)\} = \phi$ so that there is $t \geqq 0$ such that $p \in \partial\tilde{W}(t)$. Corollary 10.8 implies that there is such a t at most one. Assume that p_n converges to p and $t(p_n)$ converges to τ. By Proposition 10.7, $\tau \leqq t(p)$. If $\tau < t(p)$, then $p_n \in \partial\tilde{W}(t(p_n)) \subset \tilde{W}((\tau+t(p))/2)$ for large n. Hence, by Proposition 10.7, there is $\delta > 0$ such that $\tilde{W}((\tau+t(p))/2) \oplus \delta \subset \tilde{W}(t(p))$. Therefore $p_n \in \tilde{W}((\tau+t(p))/2) \subset (\tilde{W}((\tau+t(p))/2) \oplus \delta) \ominus \delta \subset \tilde{W}(t(p)) \ominus \delta$ so that $d(p, \partial\tilde{W}(t(p)) \geqq \delta$.

This is a contradiction. Hence $\tau = t(p)$ and t is continuous.

Corollary 10.10. Let ν be a measure as in Theorem 3.4 and let $\tilde{W}$ be the minimum domain in $Q(\nu, SL^1)$. Then $m(\partial\tilde{W}) = 0$, $[\tilde{W}] = (\bar{\tilde{W}})^\circ$ and $\partial[\tilde{W}] = \partial(\tilde{W}^e)$.

Proof. Let $\mu = \nu - \chi_W m$ and $\{\nu(t)\} = \{\chi_W m + t\mu\}$. Then, by Proposition 10.7, $\partial\tilde{W} = \partial\tilde{W}(1) \subset \tilde{W}(1+\varepsilon)\setminus\tilde{W}(1)$ for every $\varepsilon > 0$. Hence $m(\partial\tilde{W}) = 0$. Since $m(\tilde{W}) \leqq m((\bar{\tilde{W}})^\circ) \leqq m(\bar{\tilde{W}}) = m(\tilde{W})$ and $[\tilde{W}] \subset (\bar{\tilde{W}})^\circ$, we obtain $[\tilde{W}] = (\bar{\tilde{W}})^\circ$. Therefore $\partial[\tilde{W}] = \partial((\bar{\tilde{W}})^\circ) = \partial(\tilde{W}^e)$.

Proposition 10.11. Let $\{\nu(t)\}$ be continuous and let $\{\Omega(t)\} \in Q(\{\nu(t)\}, SL^1)$. Then, $\mathrm{stag}\{\Omega(t)\} = \mathrm{stag}\{[\tilde{W}(t)]\} = \mathrm{disc}\{[\tilde{W}(t)]\}$, where $\{[\tilde{W}(t)]\}$ is the maximum family in $Q(\{\nu(t)\}, SL^1)$.

Proof. By definition , if $[\{\Omega_1(t)\}] = [\{\Omega_2(t)\}]$, then $\mathrm{stag}\{\Omega_1(t)\} = \mathrm{stag}\{\Omega_2(t)\}$. Hence $\mathrm{stag}\{\Omega(t)\} = \mathrm{stag}\{[\tilde{W}(t)]\}$ for every $\{\Omega(t)\} \in Q(\{\nu(t)\}, SL^1)$. Let us use the notation D, D(t,L), etc. in the Lemma 10.1 for the class $\{[\tilde{W}(t)]\}$. By Proposition 10.4, $\{[\tilde{W}(t)]\}$ is right continuous with respect to distance so that $\mathrm{disc}\{[\tilde{W}(t)]\} = \cup_{t\in L} D(t,L) \subset \mathrm{stag}\{[\tilde{W}(t)]\}$. Let $p \in \mathrm{stag}\{[\tilde{W}(t)]\}$, and let $t(p) \geqq 0$ and $\varepsilon = \varepsilon(p) > 0$ be numbers such that $m(\Delta(r;p)\setminus[\tilde{W}(t)]) > 0$ for $r > 0$ and $t < t(p)$ and $m(\Delta(\varepsilon;p)\setminus[\tilde{W}(t)]) = 0$ for $t \geqq t(p)$. Since $p \in [\tilde{W}(t)]^e$ for every $t < t(p)$, $p \notin \cup_{t<t(p)} [\tilde{W}(t)] \in Q(\nu(t(p)), SL^1)$. On the other hand, $\Delta(\varepsilon;p) \subset [\tilde{W}(t(P))]$, because $m(\Delta(\varepsilon;p)\setminus[\tilde{W}(t(p))]) = 0$. Therefore $p \in$

$D(t(p),L)$ so that $\cup_{t\in L} D(t,L) = \text{stag}\{[W(t)]\}$.

Now let us deal with families of quadrature domains for class HL^1. For the sake of simplicity, we deal with only the families $\{\nu(t)\}$ of measures given before Proposition 10.5. At first we show

Lemma 10.12. Let $F(\Omega)$ be a subclass of $L^1(\Omega)$ such that $-f \in F(\Omega)$ for every $f \in F(\Omega)$. Then $\{\Omega(t)\} \in Q(\{\nu(t)\},F)$ if and only if $\{\Omega(t)\} \in Q(\{\chi_{\Omega(0)}m + \nu(t) - \nu(0)\},F)$ for some $\Omega(0) \in Q(\nu(0),F)$.

Proof. If $\{\Omega(t)\} \in Q(\{\nu(t)\},F)$, then $\Omega(0) \in Q(\nu(0),F)$ and

$$\int_{\Omega(0)} fdm + \int fd(\nu(t)-\nu(0)) = \int fd\nu(t) = \int_{\Omega(t)} fdm$$

for every $f \in F(\Omega(t))$. Hence $\{\Omega(t)\} \in Q(\{\chi_\Omega m + \nu(t) - \nu(0)\},F)$. The converse follows similarly.

Next let us show the following remarkable theorem.

Theorem 10.13. Let $\{\nu(t)\}$ be continuous and let $\{\Omega(t)\} \in Q(\{\nu(t)\},HL^1)$. Then $\{\Omega(t)\} \in Q(\{\chi_{\Omega(0)}m + \nu(t) - \nu(0)\},SL^1)$.

Proof. Let $\{\tilde{W}(t)\}$ be a family of domains such that $\tilde{W}(t) = \Omega(0)$ for t with $\nu(t) = \nu(0)$ and $\tilde{W}(t)$ is the minimum domain in $Q(\chi_{\Omega(0)} + \nu(t) - \nu(0),SL^1)$ for t with $\nu(t) \neq \nu(0)$. To prove the theorem, it is sufficient to show $\{\tilde{W}(t)\} \subset \{\Omega(t)\}$.

Suppose $\tilde{W}(t) \not\subset \Omega(t)$ for some $t > 0$. Then, by Corollary

10.10, $m(\partial\tilde{W}(t)) = 0$ so that $m(\Omega(t) \cap \tilde{W}(t)^e) = m(\Omega(t)\backslash\tilde{W}(t)) > 0$. Take an open disc Δ in $\Omega(t) \cap \tilde{W}(t)^e$. Since $\tilde{W}(s) \subset \tilde{W}(t)$ for s with $0 \leqq s \leqq t$, $\Delta \subset \tilde{W}(s)^e$ for such s.

We shall show that if $\Delta \subset \Omega(s)^e$ for some s with $0 \leqq s \leqq t$, then $\Delta \subset \Omega(s')^e$ for s' such that $s \leqq s'$ and $\|\nu(s')-\nu(s)\| < m(\Delta)/2$ By Corollary 4.9, $\mathrm{cap}((\partial\Omega(s')) \cap \Delta) \leqq \mathrm{cap}((\partial\Omega(s')) \cap \tilde{W}(s')^e) = 0$ so that $m(\Omega(s') \cap \Delta) = m(\Delta)$ or $\Delta \subset \Omega(s')^e$. Since $m(\Omega(s')\backslash\Omega(s)) = \|\nu(s')-\nu(s)\| < m(\Delta)/2$, $m(\Omega(s') \cap \Delta) \leqq m(\Delta)/2$ so that $\Delta \subset \Omega(s')^e$.

Since $\{\nu(t)\}$ is continuous, we can choose a sequence $\{s_j\}_{j=0}^k$ so that $0 = s_0 < s_1 < \cdots < s_k = t$ and $\|\nu(s_{j+1})-\nu(s_j)\| < m(\Delta)/2$ for $j = 0,\cdots,k-1$. Therefore $\Delta \subset \tilde{W}(0)^e = \Omega(0)^e$ implies $\Delta \subset \Omega(t)^e$. This is a contradiction.

Remark. From Lemma 10.12 and Theorem 10.13, we obtain $Q(\{\chi_{\Omega(0)}m + \nu(t) - \nu(0)\},HL^1) = Q(\{\chi_{\Omega(0)}m + \nu(t) - \nu(0)\},SL^1)$ if $Q(0) \in Q(\nu(0),HL^1)$.

Corollary 10.14. Let $\{\nu(t)\}$ be continuous and let $\{\Omega_j(t)\}$, $j = 1, 2$ be families in $Q(\{\nu(t)\},HL^1)$. Then $[\{\Omega_1(t)\}] = [\{\Omega_2(t)\}]$ if and only if $[\Omega_1(0)] = [\Omega_2(0)]$.

Finally we shall deal with quadrature domains of ν with large $\|\nu\|$. In what follows, we write $\Delta(\rho)$ for $\Delta(\rho;0)$ and assume that ν is a finite positive measure.

Lemma 10.15. Let $\Omega \in Q(\nu,SL^1)$ and $\mathrm{supp}\,\nu \subset \overline{\Delta(\rho)}$. If $\|\nu\| \geqq 36\pi\rho^2$, then $\Delta(\sqrt{\|\nu\|/\pi}/2) \subset \Omega$.

Proof. As in the proof of Lemma 3.1, we consider the function $M_{2\rho}\nu$ defined before Proposition 2.5. Then $(M_{2\rho}\nu)(z) = \|\nu\|/4\pi\rho^2$ on $\Delta(\rho)$, $(M_{2\rho}\nu)(z) = 0$ on $\Delta(3\rho)^c$ and

$$\int s\,d\nu \leqq \int_{\Delta(3\rho)} s(M_{2\rho}\nu)\,dm$$

for every $s \in SL^1(\Delta(3\rho))$. If $\|\nu\| \geqq 36\pi\rho^2$, then $3\rho \leqq \sqrt{\|\nu\|/\pi}/2$. Set

$$f(z) = \begin{cases} 1 & \text{on} \quad \Delta(\rho) \\ (M_{2\rho}\nu)(z) + 1 & \text{on} \quad \Delta(\sqrt{\|\nu\|/\pi}/2)\setminus\Delta(\rho). \end{cases}$$

Then $f(z) \geqq 1$ on $\Delta(\sqrt{\|\nu\|/\pi}/2)$ and

$$\int s\,d\nu \leqq \int_{\Delta(\sqrt{\|\nu\|/\pi}/2)} sf\,dm$$

for every $s \in SL^1(\Delta(\sqrt{\|\nu\|/\pi}/2))$. Hence $\Delta(\sqrt{\|\nu\|/\pi}/2) \subset \Omega$.

Lemma 10.16. Let $\Omega \in Q(\nu, AL^1)$ and supp $\nu \subset \overline{\Delta(\rho)}$. If $\|\nu\| \geqq 4\pi\rho^2$, then $\Delta(\sqrt{\|\nu\|/\pi}-\rho) \subset \Delta(\rho) \cup \Omega$.

Proof. For $\zeta = re^{i\varphi} \in \Delta(\rho)^c$, we have

$$(10.1) \qquad |\hat{\nu}(\zeta)| = \left|\int \frac{1}{r-e^{-i\varphi}z}\,d\nu(z)\right|$$

$$\geqq \int \operatorname{Re}\frac{1}{r-e^{-i\varphi}z}\,d\nu(z)$$

$$\geqq \frac{\|\nu\|}{r+\rho}\,.$$

By the inequality (3) given in §1,

$$\|\hat{\chi}_\Omega\|_\infty \leqq \sqrt{\pi m(\Omega)} = \sqrt{\pi\|\nu\|}.$$

Since $\hat{\nu}(\zeta) = \hat{\chi}_\Omega(\zeta)$ for $\zeta \in \Omega^c$,

$$r \geqq \sqrt{\|\nu\|/\pi} - \rho$$

for every $\zeta = re^{i\varphi} \in (\Delta(\rho) \cup \Omega)^c$. Hence $\Delta(\sqrt{\|\nu\|/\pi}-\rho) \subset \Delta(\rho) \cup \Omega$.

Lemma 10.17. Let $\Omega \in Q(\nu, AL^1)$, supp $\nu \subset \overline{\Delta(\rho)}$ and $\|\nu\| > (1+\sqrt{5})^2\pi\rho^2$. Then $\hat{\nu} \neq \hat{\chi}_\Omega$ on $\Omega\setminus\overline{\Delta(\rho)}$.

Proof. We may assume supp $\nu \subset \Delta(\rho)$. Set $G = \Omega\setminus\overline{\Delta(\rho)}$. By Lemma 10.16, $\partial\Delta(\rho)$ is an isolated boundary component of G. Let $\hat{\chi}_\Omega(z) = -\pi\bar{z} + f$, where f is a function analytic on Ω and continuou on $\bar{\Omega}$.

Let us first show that the analytic function $f - \hat{\nu}$ on G does not vanish on G. Let $\{\Omega_n\}$ be an exhaustion of Ω such that each $\partial\Omega_n$ consists of a finite number of mutually disjoint analytic simple curves. Set $G_n = \Omega_n\setminus\overline{\Delta(\rho)}$. We may assume that $\partial\Delta(\rho)$ is an isolated boundary component of G_n. Since $f - \hat{\nu}$ is continuous on $\bar{G}$ and $(f-\hat{\nu})(z) = \pi\bar{z}$ on $(\partial G)\setminus\partial\Delta(\rho)$,

$$\frac{1}{2\pi}\int_{(\partial G_n)\setminus\partial\Delta(\rho)} d\arg(f-\hat{\nu}) = -1$$

for large n. Since $\|\nu\| > (1+\sqrt{5})^2\pi\rho^2$, by the inequalities in the proof of Lemma 10.16,

(10.2) $$|\hat{\nu}(z)| - |f(z)| \geqq \frac{\|\nu\|}{2\rho} - |\hat{\chi}_\Omega(z)+\pi\bar{z}|$$

$$\geqq \frac{\|\nu\|}{2\rho} - \sqrt{\pi\|\nu\|} - \pi\rho$$

$$> \pi\rho$$

on $\partial\Delta(\rho)$.

Hence

$$\frac{1}{2\pi}\int_{-\partial\Delta(\rho)} d\arg(f-\hat{\nu}) = -\frac{1}{2\pi}\int_{\partial\Delta(\rho)} d\arg\,\hat{\nu}.$$

We also see, by (10.1), that $|\hat{\nu}(z)| \neq 0$ on $\Delta(\rho)^c$. Therefore

$$\frac{1}{2\pi}\int_{\partial\Delta(\rho)} d\arg\,\hat{\nu} = \frac{1}{2\pi}\int_{\partial\Delta(r)} d\arg\,\hat{\nu}$$

for every $r \geqq \rho$. Hence

$$\frac{1}{2\pi}\int_{\partial\Delta(\rho)} d\arg\,\hat{\nu} = -1$$

so that

$$\frac{1}{2\pi}\int_{\partial G_n} d\arg(f-\hat{\nu}) = 0$$

for large n. Therefore $f - \hat{\nu} \neq 0$ on G.

Next consider the harmonic function $h(z) = \log|f(z)-\hat{\nu}(z)| - \log|\pi\bar{z}|$ defined on G. Since, by (10.2),

$$|\hat{\nu}(z)-f(z)| \geqq |\hat{\nu}(z)| - |f(z)| > |\pi\bar{z}|$$

on $\partial\Delta(\rho)$ and $f(z) - \hat{\nu}(z) - \pi\bar{z} = \hat{\chi}_\Omega(z) - \hat{\nu}(z) = 0$ on $(\partial G)\backslash\partial\Delta(\rho)$, $h > 0$ on $\partial\Delta(\rho)$ and $h = 0$ on $(\partial G)\backslash\partial\Delta(\rho)$. Hence $h > 0$ on G so that $\hat{\nu} \neq \hat{\chi}_\Omega$ on G.

Proposition 10.18. Let $\Omega \in Q(\nu, AL^1)$, supp $\nu \subset \overline{\Delta(\rho)}$ and $\|\nu\| > (1+\sqrt{5})^2\pi\rho^2$. Then $\Delta(\rho) \cup \Omega$ is simply connected, $\partial(\Delta(\rho) \cup \Omega)$ consists of an analytic quasi-simple curve, and $n(p) = 1$ on $\partial(\Delta(\rho) \cup \Omega)$.

Proof. By Lemma 10.16, $\Delta(\sqrt{5}\rho) \subset \Delta(\rho) \cup \Omega$. Take a point $p \in \overline{\Delta(\rho)} \cap \Omega$, let δ_p be the Dirac measure at p and let $\{\tilde{\Omega}(t)\}$ be the minimum family in $Q(\{\chi_\Omega m + t\delta_p\}, SL^1)$.

Assume first that $\Delta(\rho) \cup \Omega$ is not simply connected. Take a continuous simple curve $\gamma \subset \Omega\backslash\overline{\Delta(\rho)}$ so that some boundary point of $\Delta(\rho) \cup \Omega$ is surrounded by γ. Let G be the simply connected bounded domain surrounded by γ. We may assume $G \subset \Delta(\rho)^e$.

By using the same arguments as in Proposition 9.4, we can find $t_0 \geqq 0$ such that $E = G\backslash\tilde{\Omega}(t_0) \neq \phi$ and $m(E) = 0$. Since $\tilde{\Omega}(t_0) \in Q(\nu + t_0\delta_p, AL^1)$, $(\nu + t_0\delta_p)\hat{}(\zeta) = \hat{\chi}_{\tilde{\Omega}(t_0)}(\zeta) = \hat{\chi}_{\tilde{\Omega}(t_0)\cup E}(\zeta)$ for $\zeta \in E$. This contradicts Lemma 10.17, because $\tilde{\Omega}(t_0) \cup E \in Q(\nu+t_0\delta_p, AL^1)$ and $\|\nu + t_0\delta_p\| \geqq \|\nu\| > (1+\sqrt{5})^2\pi\rho^2$. Hence $\Delta(\rho) \cup \Omega$ is simply connected.

Since $\partial(\Delta(\rho) \cup \Omega)$ is an isolated nondegenerate boundary component of Ω, by Theorem 6.7, it is an analytic quasi-simple curve. If $\partial(\Delta(\rho) \cup \Omega)$ contains an infinite number of points p with $n(p) = 2$, then $\partial(\Delta(\rho) \cup \Omega)$ is a "slit" (or an arc). This contradict that $\Delta(\rho) \cup \Omega$ is a bounded domain.

Next assume that $(\Delta(\rho) \cup \Omega)^e$ is not connected. Then $\Delta(\rho) \cup \tilde{\Omega}(t)$ is not simply connected for sufficiently small $t > 0$, because $\overline{\Delta(\rho) \cup \Omega} \subset \Delta(\rho) \cup \tilde{\Omega}(t)$ for $t > 0$. This contradicts the above consequence. Hence $n(p) = 1$ on $\partial(\Delta(p) \cup \Omega)$.

Let ν be a complex measure with compact support on $\mathbb{C}$ and let $\Omega \in Q(\nu, AL^1)$ be a simply connected domain containing supp ν. Let $w = \varphi(z)$ be a one-to-one conformal mapping from Ω onto the unit disc Δ. We define an analytic function f on $\Delta \cup \varphi(\Omega\setminus\text{supp }\nu)^*$ by

$$f(w) = \begin{cases} \varphi^{-1}(w) & \text{on} \quad \Delta \\ \overline{\{\frac{1}{\pi}(\hat{\chi}_\Omega + \pi\bar{z}) - \frac{1}{\pi}\hat{\nu}\}} \circ \varphi^{-1}(w^*) & \text{on} \quad \varphi(\Omega\setminus\text{supp }\nu)^*, \end{cases}$$

where w^* denotes the reflection point of w with respect to $\partial\Delta$. By the same argument as in Theorem 6.7, we see that f can be extended analytically onto $\mathbb{C}\setminus\varphi(\text{supp }\nu)^*$. For a point c in $\Omega\setminus$ supp ν, take φ so that $\varphi(c) = 0$. Then we obtain the following integral representation of f:

Proposition 10.19. It follows that

$$f(w) = c + \frac{w}{\pi}\int_{\psi(\text{supp}\nu)} \frac{1}{\eta - w} \frac{\eta}{\overline{\varphi^{-1\prime}(\eta^*)}} \, \overline{d\nu\circ\psi^{-1}(\eta)}$$

on $\mathbb{C}\setminus\psi(\text{supp }\nu)$, where $\psi(z) = \varphi(z)^*$.

Proof. Let $\mu = (\nu\circ\varphi^{-1})/\varphi^{-1\prime}$. Then, by (2) in §1, $\widehat{\nu\circ\varphi^{-1}} - \hat{\mu}$ can be extended analytically onto Δ. Hence

$$f(w) + \frac{1}{\pi}\overline{\hat{\mu}(w^*)} = f(w) + \frac{1}{\pi}\int_{\eta^*\in\varphi(\mathrm{supp}\nu)} \frac{1}{\overline{\eta}^*-\overline{w}^*}\,\frac{1}{\overline{\varphi^{-1'}(\eta^*)}}\,\overline{d\nu\circ\varphi^{-1}(\eta^*)}$$

$$= f(w) - \frac{1}{\pi}\int_{\eta\in\psi(\mathrm{supp}\nu)} \frac{w\eta}{\eta-w}\,\frac{1}{\overline{\varphi^{-1'}(\eta^*)}}\,\overline{d\nu\circ\psi^{-1}(\eta)}$$

can be extended analytically onto Δ^e. Since this function is analytic on $\overline{\Delta}$ and takes c when $w = 0$, it is identically equal to a constant c. This completes the proof.

Proposition 10.20. Let $\Omega \in Q(\nu, AL^1)$, supp $\nu \subset \overline{\Delta(\rho)}$ and $||\nu|| \geqq 49\pi\rho^2$. Then $\partial(\Delta(\rho) \cup \Omega)$ consists of an analytic simple curve.

Proof. We may assume $0 \in \Omega\backslash\mathrm{supp}\,\nu$. By Proposition 10.18, $\Delta(\rho) \cup \Omega$ is simply connected. By Lemma 10.16, $\Delta(6\rho) \subset \Delta(\rho) \cup \Omega$. Apply Proposition 10.19 replacing Ω by $\Delta(\rho) \cup \Omega$. Let φ be the one-to-one conformal mapping from $\Delta(\rho) \cup \Omega$ onto the unit disc Δ with $\varphi(0) = 0$ and $\varphi'(0) > 0$. Then, by Proposition 10.19, $f = \varphi^{-1}$ satisfies

$$f'(w) = \frac{1}{\pi}\int \frac{\eta^2}{(\eta-w)^2}\,\frac{1}{\overline{\varphi^{-1'}(\eta^*)}}\,d\xi(\eta)$$

on $\partial\Delta$, because $\xi = (\nu+\chi_{\Delta(\rho)\backslash\Omega}m)\circ\psi^{-1}$ is positive. Since $1/|\eta| \leqq 1/$

$$\left|\arg\frac{\eta^2}{(\eta-w)^2}\right| = 2\left|\arg\left(1-\frac{w}{\eta}\right)\right| \leqq 2 \text{ arc sin } \frac{1}{6}.$$

By the Goluzin rotation theorem,

$$\left|\arg\frac{1}{\overline{\varphi^{-1'}(\eta^*)}}\right| = \left|\arg\varphi'(\zeta)\right| \leqq 4 \text{ arc sin } \frac{1}{6},$$

where $\zeta = \varphi^{-1}(\eta^*) \in \overline{\Delta(\rho)}$. Hence

$$|\arg f'(w)| \leqq 6 \arcsin \frac{1}{6} < \frac{\pi}{2}$$

so that $f' \neq 0$ on $\partial\Delta$. Therefore the proposition follows from Proposition 10.18.

Remark. If ν is not a positive measure, then Proposition 10.20 does not necessarily hold. Let ν be a measure corresponding to the functional $g \mapsto (3\pi/2)g(0) + (\pi/2)g'(0)$ and let $f(w) = w + (1/2)w^2$. Then $f(\Delta) \in Q(\nu, AL^1)$ and $f'(-1) = 0$. For details, see Davis [9, pp. 53-54 and pp. 159-160].

Proposition 10.21. Let $\Omega \in Q(\nu, AL^1)$, supp $\nu \subset \overline{\Delta(\rho)}$ and $\|\nu\| \geqq 49\pi\rho^2$. Then $\Delta(\rho) \cup \Omega$ is the unique domain in $Q(\nu+\chi_{\Delta(\rho)\setminus\Omega}m, SL^1)$.

Proof. Let us apply Proposition 9.4. The domain $\Delta(\rho) \cup \Omega$ is in $Q(\nu+\chi_{\Delta(\rho)\setminus\Omega}m, AL^1)$ and by Lemma 10.17, $E = E(\Delta(\rho) \cup \Omega;$ $\nu+\chi_{\Delta(\rho)\setminus\Omega}m, AL^1) \subset \overline{\Delta(\rho)}$. Let $G \in Q(\nu+\chi_{\Delta(\rho)\setminus\Omega}m, SL^1)$. Then, by Lemma 10.15, $\Delta(3.5\rho) \subset G$. Hence $E\setminus G = \phi$. Therefore $G = \Delta(\rho) \cup \Omega$.

Corollary 10.22. Let supp $\nu \subset \overline{\Delta(\rho)}$ and $\|\nu\| \geqq 49\pi\rho^2$. Let Ω_j, $j = 1,2$, be domains in $Q(\nu, AL^1)$. If $\hat{\chi}_{\Delta(\rho)\setminus\Omega_1} = \hat{\chi}_{\Delta(\rho)\setminus\Omega_2}$ on $\Delta(\rho)^e$, then $\Delta(\rho) \cup \Omega_1 = \Delta(\rho) \cup \Omega_2$.

Proof. If $\hat{\chi}_{\Delta(\rho)\setminus\Omega_1} = \hat{\chi}_{\Delta(\rho)\setminus\Omega_2}$ on $\Delta(\rho)^e$, then $\Delta(\rho) \cup \Omega_1 \in$ $Q(\nu+\chi_{\Delta(\rho)\setminus\Omega_2}m, AL^1)$. Hence $\Delta(\rho) \cup \Omega_1 = \Delta(\rho) \cup \Omega_2$.

Proposition 10.23. Let ν and Ω be as in Lemma 10.17. For a fixed ρ, we have

$$\Omega \subset \Delta\left(\sqrt{\|\nu\|/\pi} + 2\rho + O(1/\sqrt{\|\nu\|})\right).$$

Proof. We may assume $0 \in \Omega\setminus\operatorname{supp}\nu$. Since $\Delta(\rho) \cup \Omega$ is a simply connected domain in $Q(\nu+\chi_{\Delta(\rho)\setminus\Omega}m, AL^1)$, we can apply Proposition 10.19, replacing Ω by $\Delta(\rho) \cup \Omega$. Let φ be a one-to-one conformal mapping from $\Delta(\rho) \cup \Omega$ onto the unit disc Δ with $\varphi(0) = 0$. Let $r = \sqrt{\|\nu\|/\pi} - \rho$. Then $\Delta(r) \subset \Delta(\rho) \cup \Omega$ by Lemma 10.16.

Hence $|\varphi(z)| \leqq |z|/r \leqq \rho/r$ on $\overline{\Delta(\rho)}$, $|\varphi'(z)| \leqq r(1-|\varphi(z)|^2)/(r^2-|z|^2)$ on $\Delta(r)$ and

$$\left|\frac{\eta}{\eta-w}\,\frac{1}{\varphi^{-1\,\prime}(\eta^*)}\right| \leqq \frac{1}{1-|\varphi(\zeta)|}\cdot|\varphi'(\zeta)|$$

$$\leqq \frac{1}{r-\rho}$$

for $w \in \overline{\Delta}$, where $\zeta = \varphi^{-1}(\eta^*) \in \overline{\Delta(\rho)}$. Therefore

$$|\varphi^{-1}(w)| \leqq \frac{\|\nu\|+\pi\rho^2}{\pi(r-\rho)}$$

$$\leqq \left(\sqrt{\|\nu\|/\pi} + \frac{\rho^2}{\sqrt{\|\nu\|/\pi}}\right)\cdot\frac{1}{1-\dfrac{2\rho}{\sqrt{\|\nu\|/\pi}}}$$

for $w \in \overline{\Delta}$. This completes the proof.

Let φ be as in the proof of Proposition 10.23. We denote by $f(w;\nu)$ the function defined before Proposition 10.19 which is equal to $\varphi^{-1}(w)$ on Δ. If $\varphi'(0) > 0$, then, for a fixed ρ and R, we see that $f(w;\nu)/\sqrt{\|\nu\|/\pi}$ converges uniformly to w on $\Delta(R)$ as $\|\nu\|$ tends to $+\infty$. More precisely we show

Proposition 10.24. Let ν and Ω be as in Lemma 10.17. For a fixed ρ and R we have

$$\left\|\frac{f(w;\nu)}{\sqrt{\|\nu\|/\pi}} - w\right\|_{\infty,\Delta(R)} = O\left(\frac{1}{\sqrt{\|\nu\|}}\right).$$

Proof. Let $\theta(z) = \varphi(z)/z$ and $r = \sqrt{\|\nu\|/\pi} - \rho$. Then $\|\theta(z)\|_{\infty,\Delta(\rho)\cup\Omega} \leqq 1/r$ and so $|\theta'(z)| \leqq 2/(r(r-\rho))$ on $\overline{\Delta(\rho)}$. Hence

$$|\varphi'(z)-\varphi'(0)| = \left|\int_0^z \theta'(z)dz+z\theta'(z)\right|$$

$$\leqq \frac{4\rho}{r(r-\rho)}$$

on $\overline{\Delta(\rho)}$. Since $\Delta(r) \subset \Delta(\rho) \cup \Omega \subset \Delta(r+3\rho+O(1/\sqrt{\|\nu\|}))$,

$$\left|\varphi'(0) - \frac{1}{\sqrt{\|\nu\|/\pi}}\right| \leqq \frac{2\rho+O(1/\sqrt{\|\nu\|})}{r(r+\rho)}.$$

Hence

$$\left|\varphi'(z) - \frac{1}{\sqrt{\|\nu\|/\pi}}\right| \leqq O\left(\frac{1}{\|\nu\|}\right)$$

on $\overline{\Delta(\rho)}$.

Take $\|\nu\|$ so large that $\rho/r < 1/R$. Since $|\varphi(z)| \leqq \rho/r$ on $\overline{\Delta(\rho)}$, $f(w;\nu)$ is analytic on $\Delta(R)$ for such ν. Let $\xi = (\nu+\chi_{\Delta(\rho)\setminus\Omega}m)\circ\psi^{-1}$ and $\zeta = \varphi^{-1}(\eta^*) \in \overline{\Delta(\rho)}$. Then ξ is positive, $|1/\eta| \leqq \rho/r$ for $\eta \in \psi(\text{supp } \nu)$,

$$\frac{f(w;\nu)}{\sqrt{\|\nu\|/\pi}} - w = \int \left\{ \frac{w}{\sqrt{\pi}\,\|\nu\|} \frac{\eta}{\eta - w} \overline{\varphi'(\zeta)} - \frac{w}{\int d\xi} \right\} d\xi(\eta)$$

and

$$\left| \frac{1}{\sqrt{\pi}\,\|\nu\|} \frac{\eta}{\eta - w} \overline{\varphi'(\zeta)} - \frac{1}{\int d\xi} \right|$$

$$\leqq \left| \frac{1}{\sqrt{\pi}\,\|\nu\|} \frac{1}{1-\frac{w}{\eta}} \left(\overline{\varphi'(\zeta)} - \frac{1}{\sqrt{\|\nu\|/\pi}} \right) \right| + \left| \frac{1}{\sqrt{\pi}\,\|\nu\|} \frac{1}{1-\frac{w}{\eta}} \frac{1}{\sqrt{\|\nu\|/\pi}} - \frac{1}{\int d\xi} \right|$$

$$= O\left(\|\nu\|^{-\frac{3}{2}} \right).$$

Hence the proposition follows.

<u>Corollary 10.25</u>. Let ν and Ω be as in Lemma 10.17. Let γ be the image of $\partial(\Delta(\rho) \cup \Omega)$ under the mapping $\sigma(z) = z/\sqrt{\|\nu\|/\pi}$. Then the curvature of γ converges uniformly to 1 as $\|\nu\|$ tends $+\infty$. In particular, $\partial(\Delta(\rho) \cup \Omega)$ is convex for a sufficiently large $\|\nu\|$.

<u>Proof</u>. Set $g(w) = (\sigma \circ f)(w) = f(w;\nu)/\sqrt{\|\nu\|/\pi}$. Then $\gamma = g(\partial\Delta)$ and the curvature $\kappa(p)$ of γ at $p \in \gamma$ is given by

$$\kappa(p) = \frac{1}{|g'(w)|} \left(\mathrm{Re}\, \frac{wg''(w)}{g'(w)} + 1 \right),$$

where $w = g^{-1}(p)$. Hence the corollary follows from Proposition 10.24, because γ is convex if and only if $\kappa(p) \geqq 0$ on γ.

§11. Quadrature domains with infinite area

In their paper [1], Aharonov and Shapiro have treated quadrature domains Ω under the hypothesis

$$\int_\Omega \frac{dm(z)}{|z|} < +\infty$$

and have not required Ω to have finite area.

In this section, we shall show that the above hypothesis can be replaced a weaker hypothesis

$$\int_{\Omega\setminus\Delta(1;0)} \frac{dm(z)}{|z|^2} < +\infty.$$

Next we shall construct quadrature domains with infinite area of a positive measure with compact support.

At first we prepare the following lemma:

Lemma 11.1. Let $v(t)$ be a nonnegative lower semicontinuous function on $[1,+\infty)$ such that $\int_1^\infty v(t)/t^2 dt < +\infty$. For every $\varepsilon > 0$, set $r(\varepsilon) = \sup\{r \geqq 1 \mid \int_1^r v(t)dt > \varepsilon r^2\}$ (if $\int_1^r v(t)dt \leqq \varepsilon r^2$ for every $r \geqq 1$, then set $r(\varepsilon) = 1$). Then

(1) $1 \leqq r(\varepsilon) < +\infty$.

(2) $r(\varepsilon)$ increases as ε decreases.

(3) $\displaystyle\int_1^{r(\varepsilon)} v(t)dt = \varepsilon r(\varepsilon)^2$ if $r(\varepsilon) > 1$.

(4) $v(r(\varepsilon)) \leqq 2\varepsilon r(\varepsilon)$ if $r(\varepsilon) > 1$.

Proof. For $\varepsilon > 0$, take $r_1 \geqq 1$ so that $\int_{r_1}^\infty v(t)/t^2 dt < \varepsilon/2$ and set $r_2 = \max\{r_1, (2\int_1^{r_1} v(t)dt/\varepsilon)^{1/2}\}$. Then

$$\int_1^r v(t)dt = \int_1^{r_1} v(t)dt + \int_{r_1}^r v(t)dt$$

$$\leq \varepsilon r_2^2/2 + r^2\int_{r_1}^{r} v(t)/t^2 dt$$

$$\leq \varepsilon r^2/2 + \varepsilon r^2/2$$

$$= \varepsilon r^2$$

for every $r \geq r_2$. Hence (1) holds. It is easy to show (2) and (3). If $r(\varepsilon) > 1$, for every $h > 0$,

$$\int_1^{r(\varepsilon)+h} v(t)dt \leq \varepsilon\left(r(\varepsilon)+h\right)^2 .$$

By (3),

$$\int_{r(\varepsilon)}^{r(\varepsilon)+h} v(t)dt \leq \varepsilon\left(2r(\varepsilon)+h\right)h$$

so that $\inf\{v(t) \mid r(\varepsilon)\leq t\leq r(\varepsilon)+h\} \leq \varepsilon(2r(\varepsilon)+h)$. Hence $v(r(\varepsilon)) \leq \liminf_{t\to r(\varepsilon)} v(t) \leq 2\varepsilon r(\varepsilon)$.

Theorem 11.2. Let ν be a complex measure with compact support. Then

$$\int_{\Omega\setminus\Delta(1;0)} \frac{dm(z)}{|z|^2} = +\infty$$

for every $\Omega \in Q_\infty(\nu, AL^1)$.

Proof. It is sufficient to show that if Ω satisfies (1) and (2) in Introduction, and $I = \int_{\Omega\setminus\Delta(1;0)} (1/|z|^2)dm(z) < +\infty$, then $m(\Omega) < +\infty$. Take $\rho > 0$ so that $\operatorname{supp} \nu \subset \Delta(\rho;0)$. Since $I < +\infty$, $(\Omega \cup \Delta(\rho;0))^c \neq \phi$. Take $z_0 \in (\Omega \cup \Delta(\rho;0))^c$ and fix it. Define $\hat{\hat{\nu}}$ and $\hat{\hat{\chi}}_\Omega$ by

$$\hat{\nu}(z) = \int \left(\frac{1}{\zeta - z} - \frac{1}{\zeta - z_0}\right) d\nu(\zeta)$$

and

$$\hat{\hat{\chi}}_\Omega(z) = \int_\Omega \left(\frac{1}{\zeta - z} - \frac{1}{\zeta - z_0}\right) dm(\zeta),$$

respectively. Since $I < +\infty$, $\frac{1}{\zeta - z} - \frac{1}{\zeta - z_0}$ belongs to $L^1(\Omega)$ for every $z \in \mathbb{C}$. Hence $\hat{\hat{\nu}}(z) = \hat{\hat{\chi}}_\Omega(z)$ for every $z \in \Omega^c$. If $z \in \Delta(\rho;0)^c$, then $z \notin \operatorname{supp} \nu$ so that $\hat{\hat{\nu}}(z) = \hat{\nu}(z) - \hat{\nu}(z_0)$ on $\Delta(\rho;0)^c$.

As an argument similar to Theorem 6.4, for $r > \rho$, we have

$$\frac{1}{2\pi i}\int_{\partial\Delta(r;0)} \hat{\hat{\nu}}(z)dz = -\int d\nu,$$

$$\frac{1}{2\pi i}\int_{\partial\Delta(r;0)} \hat{\hat{\chi}}_\Omega(z)dz = -m(\Omega \cap \Delta(r;0))$$

and

$$m(\Omega \cap \Delta(r;0)) - \int d\nu \leqq \frac{1}{2\pi}\int_{\partial\Delta(r;0)} |\hat{\hat{\chi}}_\Omega(z) - \hat{\hat{\nu}}(z)|\chi_\Omega(z) r d\theta,$$

where $z = re^{i\theta}$.

Now assume $m(\Omega) = +\infty$ and set $v(r) = \int_{\partial\Delta(r;0)} \chi_\Omega(z) r d\theta$. Since v is nonnegative and lower semicontinuous, and $I = \int_1^\infty v(t)/t^2 dt < +\infty$, by Lemma 11.1, we can find $r(\varepsilon)$ satisfying from (1) to (4) in Lemma 11.1. By the assumption, $r(\varepsilon) \uparrow +\infty$ as $\varepsilon \downarrow 0$. Take $\varepsilon > 0$ so that $\varepsilon \leqq 1/e$ and $r(\varepsilon) > \rho + 1$, and set $r = r(\varepsilon)$.

Since $v(r) \leqq 2\varepsilon r$, by using an argument similar to Theorem 6.4, we can find $\zeta = \zeta(z) \in \Omega^c \cap \partial\Delta(r;0)$ satisfying $|\zeta - z| \leqq |\int_z^\zeta r d\theta| \leqq v(r)/2 \leqq \varepsilon r$. Since

$$\hat{\hat{\chi}}_\Omega(z) - \hat{\hat{\chi}}_\Omega(\zeta) = \int_\Omega \left(\frac{1}{w-z} - \frac{1}{w-\zeta}\right) dm(w)$$

$$= (z-\zeta)\int_{\Omega\cap\Delta} \frac{dm(w)}{(w-z)(w-\zeta)} + \int_\zeta^z \left\{\int_{\Omega\setminus\Delta} \frac{dm(w)}{(w-\eta)^2}\right\} d\eta,$$

where $\Delta = \Delta(3r;(z+\zeta)/2)$, and since $|w-\eta| \geqq |w| - |\eta| \geqq |w| - r \geqq |w| - |w|/2 = |w|/2$ on Δ^c, by Lemma 6.2, we have

$$\frac{1}{2\pi}|\hat{\hat{\chi}}_\Omega(z) - \hat{\hat{\chi}}_\Omega(\zeta)| \leqq |z-\zeta|\left\{1+\log\frac{6r}{|z-\zeta|} + \frac{1}{2\pi}4I\right\}$$

$$= r\cdot\frac{|z-\zeta|}{r}\left\{\log\frac{r}{|z-\zeta|} + (1+\log 6+2I/\pi)\right\}$$

$$\leqq r\varepsilon\{\log(1/\varepsilon)+(1+\log 6+2I/\pi)\}.$$

Since $r > \rho + 1$, we have

$$|\hat{\hat{\nu}}(\zeta)-\hat{\hat{\nu}}(z)| = |\hat{\nu}(\zeta)-\hat{\nu}(z)|$$

$$= \left|\int_z^\zeta \left\{\frac{d\nu(w)}{(w-\eta)^2}\right\} d\eta\right|$$

$$\leqq \varepsilon r\|\nu\|.$$

Hence, as in Theorem 6.4,

$$\int_0^r v(t)dt - \int d\nu \leqq 2(\varepsilon r)^2\left\{\log\frac{1}{\varepsilon} + \left(1+\log 6+\frac{2I}{\pi}+\frac{\|\nu\|}{2\pi}\right)\right\}.$$

By (3) in Lemma 11.1, $\int_0^r v(t)dt - \int d\nu = \varepsilon r^2 + \int_0^1 v(t)dt - \int d\nu$. Therefore we have

$$\varepsilon r(\varepsilon)^2 + C_1 \leqq 2(\varepsilon r(\varepsilon))^2(\log\frac{1}{\varepsilon} + C_2),$$

where C_1 and C_2 are constants. Divide both sides by $\varepsilon r(\varepsilon)^2$ and let ε tend to 0. By the assumption, $\varepsilon r(\varepsilon)^2 \uparrow +\infty$ as $\varepsilon \downarrow 0$.

Hence we get $1 \leqq 0$. This is a contradiction.

Next we note that $AL^1(\mathbb{C}) = \{0\}$ and so $\mathbb{C} \in Q_\infty(\nu, AL^1)$ for every complex measure ν. More precisely we have

Lemma 11.3. Let Ω be a plane domain and let n be the cardinal number of $\Omega^c = \mathbb{C}\setminus\Omega$. Then $\dim_{\mathbb{C}} AL^1(\Omega) = \max\{n-2, 0\}$ and $\dim_{\mathbb{R}} HL^1(\Omega) = \max\{3n-5, 0\}$. For class SL^1, it follows that $SL^1(\mathbb{C}) = \{0\}$ and, for every $\Omega \neq \mathbb{C}$, there is a sequence $\{s_j\}_{j=1}^\infty \subset SL^1(\Omega)$ such that the function 0 and s_j, $j = 1, 2, \cdots$, are extreme points of the convex hull of $\{0\} \cup \{s_j\}_{j=1}^\infty$.

Proof. Let $\Omega^c = \{\zeta_1, \zeta_2, \cdots, \zeta_n\}$. Then, by Lemma 6.8, every f in $AL^1(\Omega)$ can be expressed as $f(z) = \Sigma a_j/(z-\zeta_j)$ for some constants a_j. Hence $AL^1(\Omega) = \{0\}$ if $n \leqq 2$. For $n \geqq 3$, take ζ_k, $k = 3, \cdots, n$. Then $f_k(z) = a_1/(z-\zeta_1) + a_2/(z-\zeta_2) + 1/(z-\zeta_k) \in AL^1(\Omega)$ if and only if $a_1(z-\zeta_2)(z-\zeta_k) + a_2(z-\zeta_1)(z-\zeta_k) + (z-\zeta_1)\cdot(z-\zeta_2)$ is constant, namely, if and only if $a_1 = (\zeta_2-\zeta_k)/(\zeta_1-\zeta_2)$ and $a_2 = (\zeta_1-\zeta_k)/(\zeta_2-\zeta_1)$. Therefore $\{f_k\}_{k=3}^n \subset AL^1(\Omega)$ is a basis of $AL^1(\Omega)$.

If $h \in HL^1(\Omega)$, then h can be expressed as

$$h(z) = \mathrm{Re} \sum \frac{a_j}{z-\zeta_j} + \sum \alpha_j \log|z-\zeta_j| + u(z),$$

where a_j are complex constants, α_j are real constants and u is a harmonic function on $\mathbb{C}$. Set $L(r) = \int h(re^{i\theta}+c)d\theta$ for a fixed $c \in \mathbb{C}$. Since $\int_0^\infty |L(r)|rdr \leqq \int_\Omega |h|dm < +\infty$, $\liminf_{r\to+\infty} |L(r)| = 0$. Hence $\Sigma\alpha_j = 0$ and $u(c) = 0$. Therefore $HL^1(\Omega) = \{0\}$ if $n \leqq 1$.

If $n = 2$, then

$$(11.1)\qquad h_1(z) = \frac{1}{2}\,\mathrm{Re}\left\{\frac{\zeta_1-\zeta_2}{z-\zeta_1} + \frac{\zeta_1-\zeta_2}{z-\zeta_2}\right\} + \log\left|\frac{z-\zeta_1}{z-\zeta_2}\right|$$

belongs to $HL^1(\Omega)$ and $\{h_1\}$ is a basis of $HL^1(\Omega)$, because

$$\log\left|\frac{z-\zeta_1}{z-\zeta_2}\right| = \mathrm{Re}\,\log\left(1 - \frac{\zeta_1-\zeta_2}{z-\zeta_2}\right) = -\mathrm{Re}\left\{\sum_{j=1}^{\infty} \frac{1}{j}\left(\frac{\zeta_1-\zeta_2}{z-\zeta_2}\right)^j\right\}$$

for large z. If $n \geqq 3$, we define h_k, $k = 3,\cdots,n$, replacing ζ_1 by ζ_k in (11.1). Then h_k, $\mathrm{Re}\, f_k$, $\mathrm{Re}\, if_k$, $k = 3,\cdots, n$ and h_1 form a basis of $HL^1(\Omega)$ by themselves.

Finally we deal with the class SL^1. Let $s \in SL^1(\mathbb{C})$ and set $L(r) = \int s(re^{i\theta}+c)d\theta$. Then $L(r)$ is increasing, is a convex function of $\log r$ and $\liminf_{r\to+\infty} |L(r)| = 0$. Hence $L(r) = 0$ for every $r > 0$ so that $s(c) = \lim_{r\to 0} L(r) = 0$ for every $c \in \mathbb{C}$.

Let $\Omega \neq \mathbb{C}$ and let $\zeta \in \partial\Omega$. We may assume that $\Omega \cap \{z \in \mathbb{C} \mid |z-\zeta| = 1/j\} \neq \phi$ for every natural number j. Set $s_j(z) = \max\{1/|z-\zeta|-j,0\}$, $j = 1, 2,\cdots$. Then $s_j \in SL^1(\Omega)$ for every j. Assume that $s_k = \Sigma_{j=1}^n\, \alpha_j s_j$ for nonnegative constants α_j. Then, for $z \in \Omega$ with $|z-\zeta| < \min\{1/k,1/n\}$,

$$\frac{1}{|z-\zeta|} - k = \frac{\Sigma\alpha_j}{|z-\zeta|} - \Sigma\, j\alpha_j.$$

Hence $\Sigma\, \alpha_j = 1$ and $k = \Sigma\, j\alpha_j$. Next take $z \in \Omega$ so that $|z-\zeta| = 1/k$. Then

$$0 = \sum_{j=1}^{k-1} \alpha_j\left\{\frac{1}{|z-\zeta|} - j\right\} = \sum_{j=1}^{k-1} (k-j)\alpha_j.$$

Hence $\alpha_j = 0$, $j = 1,\cdots,k-1, k+1,\cdots,n$ and $\alpha_k = 1$. This completes the proof.

Now we deal with the "null" quadrature domains.

Lemma 11.4. Let Ω be one of the following domains:

(1) The complement of a closed disc.

(2) A half plane.[1)]

(3) There is a line L on $\mathbb{C}$ such that $\partial\Omega \subsetneq L$.

Then $\Omega \in Q_\infty(0,SL^1)$.

Proof. If Ω is the complement of a closed disc, then we may assume $\Omega = \{z \in \mathbb{C} \mid |z| > 1\}$. Let $s \in SL^1(\Omega)$ and set $L(r) = \int s(re^{i\theta})d\theta$ for $r > 1$. Since $L(r)$ is a convex function of $\log r$ and $\liminf_{r\to+\infty} |L(r)| = 0$, $L(r) \geqq 0$ for every $r > 1$ so that $\int_\Omega s dm = \int_1^\infty L(r) r dr \geqq 0$. Hence $\Omega \in Q_\infty(0,SL^1)$.

If Ω is a half plane, then we may assume $\Omega = \{z \in \mathbb{C} \mid \mathrm{Re}\, z > 0\}$. Let $s \in SL^1(\Omega)$. For every $\varepsilon > 0$, take $\rho = \rho(\varepsilon) > 0$ so that $\int_{\Omega\setminus\Delta(\rho;\rho)} |s| dm < \varepsilon$. Let $H_s^{\Delta(\rho;4\rho)}$ be the solution in $\Delta(\rho;4\rho)$ of the Dirichlet problem for the boundary function s and set

$$s^*(z) = \begin{cases} H_s^{\Delta(\rho;4\rho)}(z) & \text{on} \quad \Delta(\rho;4\rho) \\ s(z) & \text{on} \quad \Omega\setminus\Delta(\rho;4\rho). \end{cases}$$

Then $s^* \in SL^1(\Omega)$ and

1) In his letter, Prof. H. S. Shapiro has written to the author that domains stated (2) and (3) are in $Q_\infty(0,AL^1)$.

$$\int_{\Delta(4\rho;4\rho)} s^*dm \geqq \pi(4\rho)^2 s^*(4\rho)$$

$$= 16\int_{\Delta(\rho;4\rho)} s^*dm$$

$$\geqq 16\int_{\Delta(\rho;4\rho)} sdm$$

$$\geqq -16\varepsilon$$

so that

$$\int_\Omega sdm = \int_{\Delta(4\rho;4\rho)} s^*dm - \int_{\Delta(\rho;4\rho)} s^*dm + \int_{\Delta(\rho;4\rho)} sdm$$

$$+ \int_{\Omega\setminus\Delta(4\rho;4\rho)} sdm$$

$$\geqq -16\varepsilon - \int_{\Delta(2\rho;4\rho)\setminus\Delta(\rho;4\rho)} |s|dm - \int_{\Delta(\rho;4\rho)} |s|dm$$

$$- \int_{\Omega\setminus\Delta(4\rho;4\rho)} |s|dm$$

$$\geqq -17\varepsilon$$

for every $\varepsilon > 0$. Hence $\int_\Omega sdm \geqq 0$ and $\Omega \in Q_\infty(0,SL^1)$.

If Ω is a domain stated in (3), then $\Omega \in Q_\infty(0,SL^1)$, because $\Omega\setminus L$ consists of two half planes. This completes the proof.

Starting from null quadrature domains we can always construct quadrature domains of positive measures.

<u>Theorem 11.5</u>. Let ν be a positive measure with compact support. Then there is a domain in $Q_\infty(\nu,SL^1)$ which is not the whole plane.

Proof. Let W be a half plane containing supp ν. Let $\{W_n\}$ be an exhaustion of W such that supp $\nu \subset W_1$ and each W_n is a bounded domain with quasi-smooth boundary. By using an argument similar to Theorem 3.7, we can construct a domain $\tilde{W}$ satisfying

$$\int s(d\nu+\chi_W dm) \leqq \int_{\tilde{W}} s dm$$

for every $s \in SL^1(\tilde{W})$. Since $W \in Q_\infty(0,SL^1)$, $\tilde{W} \in Q_\infty(\nu,SL^1)$. It is easy to show that $m(\tilde{W}\backslash W) \leqq \int d\nu$, and hence $\tilde{W} \neq \mathbb{C}$.

If W is the complement of a closed disc $\overline{\Delta}$ with $m(\overline{\Delta}) > \int d\nu$ and satisfies supp $\nu \subset W$, then we can also construct a domain $\tilde{W}$ in $Q_\infty(\nu,SL^1)$ with $\tilde{W} \neq \mathbb{C}$.

CHAPTER III. APPLICATIONS

§12. Analytic functions with finite Dirichlet integrals

In this section we shall study the extremal functions in the class of analytic functions with finite Dirichlet integrals on a Riemann surface.

Let R be a Riemann surface and let ζ be a fixed point on R. We denote by $AD(R,\zeta)$ the complex linear space of analytic functions f on R such that $f(\zeta) = 0$ and the Dirichlet integrals

$$D_R[f] = \int |f'(z)|^2 dxdy \qquad (z = x+iy)$$

of f on R are finite. An inner product on $AD(R,\zeta)$ is defined by

$$(f,g) = \frac{1}{\pi}\int_R f'(z)\overline{g'(z)}dxdy$$

for every pair of f and g in $AD(R,\zeta)$. With this inner product $AD(R,\zeta)$ becomes a Hilbert space. Set $\|f\| = (f,f)^{1/2} = (D_R[f]/\pi)^{1/2}$.

Let L be a bounded linear functional on $AD(R,\zeta)$. Then, by the Riesz theorem, there exists a unique function M_L in $AD(R,\zeta)$ such that $L(f) = (f,M_L)$ for every $f \in AD(R,\zeta)$.

Let t be a local coordinate defined in a neighborhood of ζ. Then the functional $f \mapsto (df/dt)(\zeta)$ is bounded. We denote the above function M for this functional by $M(z;\zeta,t,R)$.

Theorem 12.1 ([17]). The function $w = M(z) = M(z;\zeta,t,R)$ is bounded and satisfies $||M||_\infty \leq ||M||$.

<u>Proof</u>. By considering an exhaustion, without loss of generality we may assume that R is a subdomain of a compact Riemann surface S and the boundary ∂R in S consists of a finite number of mutually disjoint analytic simple curves. In this case, M can be extended analytically onto ∂R (cf. [19, pp. 114-137]) and so $W = M(R)$ is bounded and the valence function ν_M of M is bounded.

For every $f \in AD(W,0)$, we have $\int_W |f'|\nu_M dm \leq (\int_W |f'|^2 dm \cdot \int \nu_M^2 dm)^{1/2} < +\infty$ and so

$$f'(0)\frac{dM}{dt}(\zeta) = (f\circ M, M)$$

$$= \frac{1}{\pi}\int_R (f'\circ M)|M'|^2 dxdy$$

$$= \frac{1}{\pi}\int_W f'\nu_M dm$$

for every $f \in AD(W,0)$.

By Theorems 3.7 and 6.4, we can find a bounded domain $\tilde{W}$ satisfying

$$\int_W f'\nu_M dm = \int_{\tilde{W}} f' dm$$

for every $f \in AD(\tilde{W},0)$. Since $(dM/dt)(\zeta) = (M,M) = \int \nu_M dm/\pi = m(\tilde{W})/\pi$,

$$f'(0) = \frac{1}{m(\tilde{W})}\int_{\tilde{W}} f' dm$$

for every $f \in AD(\tilde{W},0)$. Hence $M(w;0,w,\tilde{W}) = (\pi/m(\tilde{W}))w$ and so $M(w;0,w,\tilde{W})$ is a linear function. By Theorem 2 in [16], we see that $[\tilde{W}]$ is the disc with radius $(m(\tilde{W})/\pi)^{1/2}$ and center at 0. Therefore

$$\|M\|_\infty = \sup_{w\in W} |w| \le (m(\tilde{W})/\pi)^{1/2} = \|M\|.$$

Next we deal with the "span metric". The span $S(\zeta)$ at $\zeta \in R$ is defined by

$$S(\zeta) = \frac{d}{dt} M(\zeta;\zeta,t,R).$$

The span $S(\zeta)$ depends on the choice of the local coordinate t. We denote by N_R the set of points at which the spans vanish. The metric $\sqrt{S(\zeta)}|dt|$ on $R\backslash N_R$ is called the span metric. Let $K(\zeta)$ be the Gaussian curvature of the span metric, namely,

$$K(\zeta) = -\frac{2}{S(\zeta)} \frac{\partial^2}{\partial t\partial \bar{t}} \log S(\zeta).$$

<u>Proposition 12.2</u> ([18]). It follows that $K(\zeta) \le -4$ for every ζ in $R\backslash N_R$.

<u>Proof</u>. We may assume that R is a surface stated as in the proof of Theorem 12.1. For a fixed local coordinate t defined around a fixed point ζ in $R\backslash N_R$, we consider the following extremal problems:

(1) Minimize $D_R[f]$ in the class $\{f \in AD(R,\zeta) | (df/dt)(\zeta) = 1$

(2) Minimize $D_R[f]$ in the class $\{f \in AD(R,\zeta) | (df/dt)(\zeta) = 0$ and $(d^2f/dt^2)(\zeta) = 1\}$.

Since $\zeta \in R \setminus N_R$, the class in (1) is not empty and there exists a unique extremal function in the class. We denote it by F_1. Since $F_1(z) = S(\zeta)^{-1}M(z;\zeta,t,R)$, by Theorem 12.1, $D_R[F_1^2/2] < +\infty$ so that the class in (2) is not empty. We denote the unique extremal function in the class by F_2.

It is known that

$$K(\zeta) = -\frac{2}{\pi}\frac{D_R[F_1]^2}{D_R[F_2]}$$

(cf. e.g. [4, Chapter III]). By definition,

$$D_R[F_2] \leq D_R[F_1^2/2].$$

Set $\varphi(w) = w^2/2$ and $W = F_1(R)$, and let ν_{F_1} be the valence function of F_1. Then $F_1^2/2 = \varphi \circ F_1$ and

$$D_R[F_1^2/2] = \int_R |(\varphi' \circ F_1)F_1'|^2 dxdy$$

$$= \int_W |\varphi'|^2 \nu_{F_1} dm.$$

Since $|\varphi'(w)|^2 = |w|^2$ is subharmonic, by Theorem 3.7, we have

$$\int_W |\varphi'|^2 \nu_{F_1} dm \leq \int_{\tilde{W}} |w|^2 dm,$$

where $\tilde{W}$ is the minimum domain in $Q(\nu_{F_1} m, SL^1)$. In Theorem 12.1 we have proved that $[\tilde{W}]$ is equal to the disc $\Delta((m(\tilde{W})/\pi)^{1/2};0)$. Hence

$$\int_{\tilde{W}} |w|^2 dm = m(\tilde{W})^2/2\pi = D_R[F_1]^2/2\pi.$$

Therefore $D_R[F_2] \leq D_R[F_1^2/2] \leq D_R[F_1]^2/2\pi$ so that $K(\zeta) \leq -4$.

For more details concerning Theorem 12.1 and Proposition 12.2, see [17] and [18].

Finally we again apply Theorem 3.7 and show a sufficient condition on the existence of quadrature domains of some complex measures. In what follows we deal with the class of analytic L^2-functions on a plane domain W with single-valued integrals. We denote it by $A'L^2(W)$. This is nothing else but the class $\{F' \mid F \in AD(W,\zeta)\}$ for some ζ in W.

We can apply our method to complex measures stated as after Theorem 6.7, but here we deal with only the following special case:

<u>Proposition 12.3</u>. Let ν be a measure corresponding to the functional $g \longmapsto a_0 g(0) + a_1 g'(0)$. If $a_0 > 0$ and $|a_1| \leq \sqrt{2/\pi}(a_0/3)^{3/2}$, then $Q(\nu, A'L^2) \neq \phi$.

<u>Proof</u>. Let L be the functional on $AD(\Delta(1;0),0)$ defined by $L(F) = \bar{b}_1 F'(0) + \bar{b}_2 F''(0)$. Then $z = M(w) = M_L(w) = b_1 w + b_2 w^2$. Let $W = M(\Delta(1;0))$ and ν_M be the valence function of M. Then, for every G in $AD(W,0)$,

$$L(G\circ M) = \bar{b}_1 (G\circ M)'(0) + \bar{b}_2 (G\circ M)''(0)$$

$$= (|b_1|^2 + 2|b_2|^2) G'(0) + b_1^2 \bar{b}_2 G''(0)$$

and

$$L(G\circ M) = \frac{1}{\pi}\int_{\Delta(1;0)} (G\circ M)'\overline{M'}\,dudv \qquad (w = u+iv)$$

$$= \frac{1}{\pi}\int_W G'\nu_M dm.$$

Apply here Theorem 3.7 and find a domain $\tilde{W}$ satisfying

$$\pi(|b_1|^2+2|b_2|^2)G'(0) + \pi b_1^2\overline{b_2}G''(0) = \int_{\tilde{W}} G'dm$$

for every $G \in AD(\tilde{W},0)$. If $a_0 > 0$ and $|a_1| \le \sqrt{2/\pi}(a_0/3)^{3/2}$, then we can find b_1 and b_2 so that $(b_1,b_2) \ne (0,0)$, $a_0 = \pi(|b_1|^2+2|b_2|^2)$ and $a_1 = \pi b_1^2\overline{b_2}$. Hence $\tilde{W} \in Q(\nu,A'L^2)$.

Aharonov and Shapiro [1] proved that if $\Omega \in Q(\nu,AL')$, then Ω is simply connected. Hence the one-to-one conformal mapping f from the unit disc onto Ω satisfying $f(0) = 0$ and $f'(0) > 0$ is a polynomial of degree 2, see before Proposition 10.19. We write it by $f(w) = b_1w + b_2w^2$, where $|b_2/b_1| \le 1/2$. By Davis [9, pp. 159-160], we have $a_0 = \pi(b_1^2+2|b_2|^2)$ and $a_1 = \pi b_1^2\overline{b_2}$ as above. If $a_0 > 0$ and $|a_1| \le \sqrt{2/\pi}(a_0/3)^{3/2}$, then we can find b_1 and b_2 satisfying $b_1 > 0$, $|b_2/b_1| \le 1/2$, $a_0 = \pi(b_1^2+2|b_2|^2)$ and $|a_1| \le \sqrt{2/\pi}(a_0/3)^{3/2}$. Hence $Q(\nu,AL^1) \ne \phi$. But our proof is simple and applicable more complicated functionals.

§13. Hele-Shaw flows with a free boundary

In this section we shall deal with Hele-Shaw flows with a free boundary produced by the injection of fluid into the narrow gap between two parallel planes.

The mathematical formulation has been given by Richardson [15]. Take Cartesian coordinates (x_1, x_2, x_3) so that the x_3-axis is perpendicular to the planes. Set $z = x_1 + ix_2$ and let $\{x_3 = 0\}$ and $\{x_3 = d\}$ $(d > 0)$ be the two planes. At some initial instant, let $\Omega(0)$ be the domain obtained by the projection of the initial blob of Newtonian fluid into the z-plane $\{x_3 = 0\}$. Take a point (a_1, a_2, d) on $\{x_3 = d\}$ so that $c = a_1 + ia_2 \in \Omega(0)$ and inject further fluid into the gap between two parallel planes.

By the theory of Hele-Shaw cell (see e.g. Lamb [14, p. 582]), for a sufficiently small $d > 0$, we obtain

$$(\overline{u}_1, \overline{u}_2) = -\frac{d^2}{12\mu}\left(\frac{\partial p}{\partial x_1}, \frac{\partial p}{\partial x_2}\right), \tag{13.1}$$

where μ denotes the coefficient of viscosity, p denotes the pressure and $\overline{u}_j$ denotes the averaged velocity. We have assumed that $u_3 = 0$ and $\frac{\partial p}{\partial x_3} = 0$, where (u_1, u_2, u_3) denotes the velocity of the fluid. The averaged velocity $\overline{u_j}$ is defined by $\overline{u_j}(x_1, x_2) = (1/d)\int_0^d u_j(x_1, x_2, x_3)dx_3$.

Hence $-(d^2/(12\mu))p$ is a potential for the averaged velocity. It is natural to assume that $p = 0$ on a free boundary of the fluid and the potential has a logarithmic singularity at c. Therefore we may assume that $-(d^2/(12\mu))p$ is a constant multiple of the Green function $g(z;c,\Omega(t))$ on $\Omega(t)$ with pole at c, where $\Omega(t)$ denotes the domain obtained by the projection of the averaged blob of the fluid at time t into the plane. By (13.1), we obtain

$$\frac{1}{\frac{\partial t}{\partial n_z}} = -\text{const.}\frac{\partial g(z;c,\Omega(t))}{\partial n_z}$$

at z on the free boundary $\partial\Omega(t)$, where $\partial/\partial n_z$ denotes the outer normal derivative at z with respect to $\Omega(t)$. In what follows, for the sake of simplicity, we take $1/(2\pi)$ for the constant in the above equation.

Now let us define rigorously a solution of Hele-Shaw flows with a free boundary. Let $\Omega(0)$ be the initial domain in $\mathbb{C}$ such that $\Omega(0) = [\Omega(0)]$, $\Omega(0)$ is bounded and $\partial\Omega(0)$ consists of a finite number of disjoint piecewise smooth quasi-simple curves. Let c be a fixed point in $\Omega(0)$. We call $\{\Omega(t)\}_{t\geq 0}$ a solution of a free boundary problem of Hele-Shaw flows with the initial domain $\Omega(0)$ (or simply, a solution of Hele-Shaw flows) if

(1) $\{\Omega(t)\}$ is monotone increasing.

(2) disc $\{\Omega(t)\} = \phi$ (for the definition, see §10).

(3) For every $t > 0$, $\Omega(t)$ is a finitely connected bounded domain such that $[\Omega(t)]\backslash\Omega(t)$ is a finite set and $\partial[\Omega(t)]$ consists of disjoint piecewise smooth quasi-simple curves.

(4) Let $D(t)$ be the set of all degenerate boundary points of $\Omega(t)$ and let $C_1(t)$ (resp. $C_2(t)$) be the set of all points z on nondegenerate boundary components of $\Omega(t)$ satisfying $n(z) = 1$ and angle $V_1 \neq \pi$ (resp. $n(z) = 2$) (for the definition, see §2). Set $E(t) = D(t) \cup C_1(t) \cup C_2(t)$, $D = \cup D(t)$, $C_j = \cup C_j(t)$ and $E = \cup E(t)$. Then $m(E) = 0$.

(5) For every $z \in \mathbb{C}\backslash\Omega(0)\backslash C_1$, there exists a unique t =

$t(z) \geq 0$ such that $z \in \partial\Omega(t)$. For every fixed $t > 0$ and every $z_0 \in (\partial\Omega(t))\setminus E(t)$, $t(z)$ is of class C^1 in a neighborhood of z_0 and $(\partial t/\partial n_{z_0})(z_0) > 0$. For every $z \in C_1(t)$ with angle $V_1 < \pi$, there is a neighborhood U of z such that $U \cap \partial\Omega(s) = \phi$ or $U \cap \partial\Omega(s)$ is connected for every $s > t$.

(6) For almost all t and every $z \in (\partial\Omega(t))\setminus E(t)$,

$$-\frac{1}{2\pi}\,\frac{\partial g(z;c,\Omega(t))}{\partial n_z}\,\frac{\partial t(z)}{\partial n_z} = 1.$$

Next we shall define a weak solution of Hele-Shaw flows. Let $\Omega(0)$ be an arbitrary bounded domain and let c be a point in $\Omega(0)$. We call $\{\Omega(t)\}$ a weak solution of Hele-Shaw flows if $\{\Omega(t)\}$ is the minimum family in $Q(\{\chi_{\Omega(0)}m+t\delta_c\},SL^1)$, where δ_c denotes the Dirac measure at c.

By using an argument similar to Theorem 3.7, we see that the family exists and is uniquely determined. Therefore a weak solution of Hele-Shaw flows always exists and is uniquely determined.

<u>Proposition 13.1</u>. A solution of Hele-Shaw flows is a weak solution.

<u>Proof</u>. Let $\{\Omega(t)\}$ be a solution of Hele-Shaw flows. For a fixed $\tau > 0$, let h be a harmonic function on $\overline{\Omega(\tau)}$. Then, by (1), (4) and (5),

$$\int_{\Omega(\tau)\setminus\Omega(0)} h\,dm = \int_{\Omega(\tau)\setminus\Omega(0)\setminus E} h(z)\,dn_z\,ds_z,$$

where ds_z denotes the line element of $\partial\Omega(t(z))$. Since $dt(z) = \frac{\partial t(z)}{\partial n_z} dn_z + \frac{\partial t(z)}{\partial s_z} ds_z = \frac{\partial t(z)}{\partial n_z} dn_z$ and $h \in L^1(\Omega(\tau))$, by using (4) and (5), we have

$$\int_{\Omega(\tau)\setminus\Omega(0)\setminus E} h(z)dn_z ds_z$$

$$= \int_0^\tau \left\{\int_{(\partial\Omega(t))\setminus E(t)} h(z) \frac{1}{\frac{\partial t(z)}{\partial n_z}} ds_z\right\} dt.$$

By (3) and (6), we obtain

$$\int_{(\partial\Omega(t))\setminus E(t)} h(z) \frac{1}{\frac{\partial t(z)}{\partial n_z}} ds_z$$

$$= \int_{(\partial\Omega(t))\setminus E(t)} h(z)\cdot -\frac{1}{2\pi} \frac{\partial g(z;c,\Omega(t))}{\partial n_z} ds_z$$

$$= h(c)$$

for almost all $t > 0$. Hence, for every $\tau \geq 0$,

(13.2) $$\int_{\Omega(0)} hdm + \tau h(c) = \int_{\Omega(t)} hdm$$

for every harmonic function on $\overline{\Omega(\tau)}$. Since the boundary of $[\Omega(\tau)]$ consists of a finite number of disjoint piecewise smooth quasi-simple curves, by Lemma 7.3, we see that (13.2) holds for every $HL^1([\Omega(\tau)])$. Hence $\{[\Omega(t)]\} \in Q(\{\chi_{\Omega(0)}m+t\delta_c\},HL^1)$.

The initial domain $\Omega(0)$ may not be a domain with quasi-smooth boundary, but by using an argument similar to Theorem 10.13, we have $Q(\{\chi_{\Omega(0)}m+t\delta_c\},HL^1) = Q(\{\chi_{\Omega(0)}m+t\delta_c\},SL^1)$, because $\chi_{\Omega(0)}m + t\delta_c = \chi_{\Omega(0)}m + (\chi_{\Omega(0)}m+t\delta_c) - \chi_{\Omega(0)}m$. Let $\{\tilde{W}(t)\}$ be the minimum family in $Q(\{X_{\Omega(0)}m+t\delta_c\},SL^1)$ and let $z \in \Omega(t)\backslash\tilde{W}(t)$ for some $t > 0$. Then $z \in [\tilde{W}(s)]^e$ for every $s < t$. Since $[\Omega(s)] = [\tilde{W}(s)]$, this implies $z \in \mathrm{disc}\{\Omega(t)\}$. By (2), it follows that $\Omega(t) = \tilde{W}(t)$ for every $t \geq 0$.

Corollary 13.2. There is at most one solution of Hele-Shaw flows.

Corollary 13.3. Let $\{\Omega(t)\}$ be a solution of Hele-Shaw flows. Then $\mathrm{stat}\{\Omega(t)\} \subset \{z \in C_1(0) \mid \text{angle } V_1 \leq \pi/2\} \subset C_1(0)$, $\mathrm{stag}\{\Omega(t)\} = D$, D and C_2 are both at most countably infinite and every nondegenerate boundary component of $\Omega(t)$ with $t > 0$ is analytic except points in $\{z \in C_1(0) \mid \text{angle } V_1 \leq \pi/2\} \cup \{z \in C_1(t) \mid \text{angle } V_1 = 2\pi\}$.

Proof. By (3) and Proposition 10.11, $\mathrm{stag}\{\Omega(t)\} = \cup D(t)$ is at most a countably infinite set. By using an argument similar to Lemma 10.2, we can construct a one-to-one mapping of C_2 onto $\mathrm{stag}\{\Omega(t)\}$. Hence C_2 is also at most a countably infinite set.

Proposition 13.4. Let $\Omega(0)$ be a domain such that $\Omega(0) = [\Omega(0)]$, $\Omega(0)$ is bounded and $\partial\Omega(0)$ consists of a finite number of disjoint piecewise smooth quasi-simple curves and let $\{\Omega(t)\}$ be a weak solution of Hele-Shaw flows with the initial domain $\Omega(0)$. If $\{\Omega(t)\}$ satisfies from (3) to (5), then it is a solution of Hele-Shaw flows.

<u>Proof</u>. Condition (1) follows from definition and (2) follows from Proposition 10.6.

We shall show (6). Fix $\tau > 0$ and let G_j, $j = 1,\cdots,n$, be the connected component of $\Omega(\tau)^e$. Then, by (5), we see that, for a sufficiently small number $\varepsilon > 0$, $\overline{G_j} \cap \partial\Omega(\tau+\varepsilon)$ is connected for every j. Take $\delta > 0$ so small that $G_j \cap (\Omega(\tau)\oplus\delta)^c \neq \phi$ for every j. By (3) and Theorem 6.4, $\overline{\Omega(\tau+\varepsilon)} \subset \Omega(\tau) \oplus \delta$ for a sufficiently small number $\varepsilon > 0$. Therefore we can find $\varepsilon_0 > 0$ so that, for every ε with $0 \leq \varepsilon \leq \varepsilon_0$, $\overline{\Omega(\tau+\varepsilon)} \subset \Omega(\tau) \oplus \delta$ and $(\Omega(\tau)\oplus\delta)^c$ intersects every connected component of $\Omega(\tau+\varepsilon)^e$.

For every j, take one point ζ_j in $G_j \cap (\Omega(\tau)\oplus\delta)^c$. Then, by the Runge theorem, the class of linear combinations of functions 1, $\log|z-\zeta_j|$, $\mathrm{Re}(z-\zeta_j)^{-k}$ and $\mathrm{Im}(z-\zeta_j)^{-k}$, $j = 1,\cdots,n$, $k = 1,2,\cdots$, are uniformly dense in the class $HC(\Omega(\tau+\varepsilon))$ of functions harmonic on $\Omega(\tau+\varepsilon)$ and continuous on $\overline{\Omega(\tau+\varepsilon)}$ for every ε with $0 \leq \varepsilon \leq \varepsilon_0$. We rearrange the above harmonic functions and denote by h_i, $i = 1,2,\cdots$.

Since $h_i \in L^1(\Omega(\tau+\varepsilon_0))$, by (4) and (5), we have

$$\int_\tau^{\tau+\varepsilon}\left\{\int_{(\partial\Omega(t))\setminus E(t)} h_i \frac{1}{\frac{\partial t(z)}{\partial n_z}}\, ds_z\right\} dt$$

$$= \int_{\Omega(\tau+\varepsilon)\setminus\Omega(\tau)} h_i\, dn_z ds_z$$

$$= \varepsilon h_i(c)$$

for every ε with $0 \leq \varepsilon \leq \varepsilon_0$. Differentiating both sides with respect to ε, we obtain

$$\int_{(\partial\Omega(t))\setminus E(t)} h_i \frac{1}{\frac{\partial t(z)}{\partial n_z}} ds_z = h_i(c)$$

a.e. on $[\tau,\tau+\varepsilon_0]$. Therefore

$$\int_{(\partial\Omega(t))\setminus E(t)} h \frac{1}{\frac{\partial t(z)}{\partial n_z}} ds_z = h(c)$$

for every $h \in HC(\Omega(t))$ for almost all t in $[\tau,\tau+\varepsilon_0]$. Since $\partial g(z;c,\Omega(t))/\partial n_z$ is continuous on $(\partial\Omega(t))\setminus E(t)$, this implies (6).

Proposition 13.5. Let $\{\Omega(t)\}$ be a solution of Hele-Shaw flows. Then, for large t, $\Omega(t)$ is a simply connected domain surrounded by an analytic simple curve. The sequence $\{\Omega(t)\}$ converges to a disc in the sense that $(\sqrt{\pi/t})(f(w;t)-f(0;t))$ converges uniformly to w on $\overline{\Delta(1;0)}$ as t tends to $+\infty$, where $f(w;t)$ is the conformal mapping from $\Delta(1;0)$ onto $\Omega(t)$ with $f(0;t) \in \Omega(0)$ and $f'(0;t) > 0$.

Proof. The proposition follows immediatly from Proposition 13.1, Lemma 10.15, Proposition 10.20 and Proposition 10.24.

Replacing $g(z;c,\Omega(t))$ in (6) by $\Sigma\alpha_j g(z;c_j,\Omega(t))$ with $\alpha_i > 0$, $\Sigma\alpha_j = 1$ and $c_j \in \Omega(0)$, we can also treat the case of a finite number of injection points c_j with different volume input rates. Its weak solution is the minimum family in $Q(\{\chi_{\Omega(0)}m+t(\Sigma\alpha_j\delta_{c_j})\},SL^1)$. Propositions 13.1, 13.4 and 13.5 also hold for this case.

Finally we give some remarks. It is plausible that a weak solution of Hele-Shaw flows is a solution. By Proposition 13.4, it is enough to show that a weak solution with an initial domain stated as in Proposition 13.4 satisfies from (3) to (5).

By using the variational inequality, we can define another weak solution of Hele-Shaw flows. The weak solution exists and is uniquely determined. Thus we have two weak solutions each of which is determined uniquely. It seems that these two weak solutions are closely related with each other. Further studies will be found in the next paper.

§14. Quadrature formulas

In 1965 Davis [6] showed that

$$\int_{-1}^{1} f(x)w(x,\rho)\,dx = \int_{E(\rho)} f\,dm$$

for every $f \in A(\overline{E(\rho)})$, where $w(x,\rho)$ is the weight function defined by $w(x,\rho) = (1/2)(\rho^2-\rho^{-2})\sqrt{1-x^2}$ for a fixed $\rho > 1$ and $E(\rho)$ denotes the ellipse defined by $E(\rho) = \{x+iy \in \mathbb{C} \mid 4x^2/(\rho+\rho^{-1})^2 + 4y^2/(\rho-\rho^{-1})^2 < 1\}$. Let $d\omega(\rho) = w(x,\rho)dx$. Then, by Lemma 7.1, we see that $E(\rho) \in Q(\omega(\rho), AL^1)$, namely, $E(\rho)$ is a quadrature domain of $\omega(\rho)$. Now, apply here the Vitali covering theorem and let $\{\Delta(r_j;z_j)\}$ be a sequence of disjoint open discs such that $\Delta(r_j;z_j) \subset E(\rho)$ and $m(E(\rho)\setminus \cup\Delta(r_j;z_j)) = 0$. Then

$$\int_{-1}^{1} f(x)w(x,\rho)\,dx = \sum a_j f(z_j) \qquad (a_j = \pi r_j^2)$$

for every $f \in AL^1(E(\rho))$. This formula was called by him a simple quadrature for class AL^1. The points z_j are distinct and the coefficients a_j and points z_j are independent of the choice of the function in $AL^1(E(\rho))$. Simple quadratures are modifications of quadratures of the form

$$\int_{-1}^{1} f(x)w(x)dx = \lim_{n\to\infty} \sum_{j=1}^{n} a_{jn} f(z_{jn})$$

which have been frequently investigated (see, Szegö [21]).

In 1969, he constructed another simple quadrature formula which has been written in Example 9.6. Further he constructed many quadrature domains, see [8] and [9].

To decide such a domain, he made use of the Bergman reproducing kernel function and the Schwarz function (see [8] and [9]). He assumed that the domain is simply connected and, for the case in Example 9.6, symmetric with respect to the real axis.

The main purpose of this section is to show that, for a positive measure on the real axis, there is an essentially unique quadrature domain for class AL^1 and the maximum domain is simply connected and symmetric with respect to the real axis.

In this section, we deal with not only domains but also open sets Ω satisfying from (1) to (3) in Introduction. We also denote by $Q(\nu,F)$ the class of all such quadrature open sets. Let Ω be an open set in $Q(\nu,AL^1)$. Then $G \in Q(\nu \mid G,AL^1)$ for every connected component G of Ω, because the function g defined by $g = f$ on G for $f \in AL^1(G)$ and $g = 0$ on $\Omega\setminus G$ belongs to $AL^1(\Omega)$. In particular, $\nu(G) = m(G) > 0$ for every G. The same holds quadrature domains for classes HL^1 and SL^1.

In what follows we denote by $\mathbb{R}$ the real axis.

<u>Lemma 14.1</u>. Let ν be a finite positive measure on $\mathbb{R}$ with compact support K such that $\inf_{z\in K}\lambda(z;\nu,100e) > 0$. Then the minimum open set in $Q(\nu,SL^1)$ is symmetric with respect to $\mathbb{R}$ and contains K in it.

<u>Proof</u>. The existence and uniqueness of the minimum open set in $Q(\nu,SL^1)$ are given by an argument similar to Theorems 3.4 and 3.5. The set is given as the union of an increasing family of open sets W_n such that each W_n is symmetric with respect to $\mathbb{R}$ and satisfies $K \subset W_1 \subset W_n \subset \overline{W}_n \subset W_{n+1}$. Hence the lemma follows.

<u>Lemma 14.2</u>. Let ν be a finite positive measure on $\mathbb{R}$ with compact support. Let Ω be an open set in $Q(\nu,AL^1)$. If Ω is symmetric with respect to $\mathbb{R}$, then $\hat{\nu} \neq \hat{\chi}_\Omega$ on $\Omega\setminus\mathbb{R}$.

<u>Proof</u>. Let G be a connected component of Ω. Since $\operatorname{Im}\hat{\nu}$ and $\operatorname{Im}\hat{\chi}_\Omega$ are both harmonic on $U = G \cap \{z \in \mathbb{C} | \operatorname{Im} z > 0\}$, $\operatorname{Im}\hat{\nu} = \operatorname{Im}\hat{\chi}_\Omega$ on $(\partial G) \cap \{z \in \mathbb{C} | \operatorname{Im} z > 0\}$, $\operatorname{Im}\hat{\chi}_\Omega = 0$ on $G \cap \mathbb{R}$, $\operatorname{Im}\hat{\nu} = 0$ on $(G \cap \mathbb{R})\setminus\operatorname{supp}\nu$ and $\operatorname{Im}\hat{\nu} > 0$ on U, we have

$$\liminf_{z\in U\to\zeta\in\partial U} \{\operatorname{Im}\hat{\nu}(z) - \operatorname{Im}\hat{\chi}_\Omega(z)\} \geq 0$$

so that $\operatorname{Im}\hat{\nu} \geq \operatorname{Im}\hat{\chi}_\Omega$ on U. If $\operatorname{Im}\hat{\nu}(z) = \operatorname{Im}\hat{\chi}_\Omega(z)$ for some $z \in U$, then $\operatorname{Im}\hat{\nu} = \operatorname{Im}\hat{\chi}_\Omega$ on U so that $\hat{\nu}$ can be extended analytically onto $G \cap \operatorname{supp}\nu$. Hence $\nu = 0$ on G. This is a contradiction and so Im $\hat{\nu} > \text{Im } \hat{\chi}_\Omega$ on U.

By the same argument, we obtain Im $\hat{\nu} < \text{Im } \hat{\chi}_\Omega$ on $G \cap \{z \in \mathbb{C} | \text{ Im } z < 0\}$ so that $\hat{\nu} \neq \hat{\chi}_\Omega$ on $\Omega \setminus \mathbb{R}$.

<u>Lemma 14.3</u>. Let ν and Ω be as in Lemma 14.1 and let G be a connected component of Ω. Then [G] is a simply connected domain surrounded by an analytic simple curve.

<u>Proof</u>. Let μ be the measure defined by $d\mu = \chi_I dx$, where I denotes a finite interval on $\mathbb{R}$ containing $G \cap \mathbb{R}$. Let $\{\tilde{G}(t)\}$ be the minimum family in $Q(\{\chi_G m + t\mu\}, SL^1)$. Since $\tilde{G}(t)$ is the minimum domain in $Q(\nu|G + t\mu, SL^1)$, by Lemma 14.1, $\tilde{G}(t)$ is symmetric with respect to $\mathbb{R}$ and contains $\bar{I}$ if $t > 0$.

Assume that [G] is not simply connected. Then there is a bounded component E of [G] such that $F = E \cap \{z \in \mathbb{C} | \text{ Im } z > 0\} \neq \phi$. If $m(F) = 0$, then $(\nu|[G])^\wedge(\zeta) = (\nu|G)^\wedge(\zeta) = \hat{\chi}_G(\zeta) = \hat{\chi}_{[G]}(\zeta)$ for $\zeta \in F \subset [G] \setminus \mathbb{R}$. This contradicts Lemma 14.2. If $m(F) > 0$, take $t_0 = m(F)/2$. Then $\tilde{G}(t_0) \supset \bar{I}$ so that we can find a continuous simple curve $\gamma \subset \tilde{G}(t_0) \cap \{z \in \mathbb{C} | \text{ Im } z > 0\}$ surrounding a simply connected domain V such that $m(V \setminus \tilde{G}(t_0)) \geq m(F)/2$. By using an argument similar to Proposition 10.18, we can choose $t_1 > t_0$ so that $V \setminus \tilde{G}(t_1) \neq \phi$ and $m(V \setminus \tilde{G}(t_1)) = 0$. Hence $(\nu|G + t_1\mu)^\wedge(\zeta) = \hat{\chi}_{\tilde{G}(t_1)}(\zeta)$ for $\zeta \in V \setminus \tilde{G}(t_1)$. This again contradicts Lemma 14.2. Thus [G] is simply connected and, by Theorem 6.7, $\partial[G]$ is an analytic quasi-simple curve.

We again use an argument similar to Proposition 10.18 and

see that $n(p) = 1$ on $\partial[G]$.

Let φ be a one-to-one conformal mapping from $[G]$ onto the unit disc Δ satisfying $\varphi(\zeta_0) = 0$ and $\varphi'(\zeta_0) > 0$ for some fixed point $\zeta_0 \in ([G] \cap \mathbb{R})\setminus \operatorname{supp} \nu$. Then $\varphi'(\zeta_0) > 0$ on $[G] \cap \mathbb{R}$, because $[G]$ is symmetric with respect to $\mathbb{R}$. Hence, by Proposition 10.19,

$$(14.1) \qquad \varphi^{-1'}(w) = \frac{1}{\pi}\int \frac{\eta^2}{(\eta-w)^2} \frac{1}{\varphi^{-1'}(\eta^*)} \, d\mu\circ\psi^{-1}(\eta),$$

where $\psi(z) = \varphi(z)^*$ and $\mu = \nu|[G]$. Since

$$\operatorname{Im} \frac{\eta^2}{(\eta-w)^2} = \frac{1}{|1-\frac{w}{\eta}|^4} \operatorname{Im}(1-\frac{\bar{w}}{\eta})^2 \gtrless 0$$

if $\eta > 1$, $|w| = 1$ and $\operatorname{Im} w \gtrless 0$, by taking ζ_0 so that $\zeta_0 < \inf_{[G]\cap \operatorname{supp} \nu} \zeta$, we see

$$\operatorname{Im} \varphi^{-1'}(w) \gtrless 0$$

for w with $|w| = 1$ and $\operatorname{Im} w \gtrless 0$ so that $\varphi^{-1'}(w) \neq 0$ if $|w| = 1$ and $w \neq \pm 1$. If $w = \pm 1$, then $\varphi^{-1'}(w) = \operatorname{Re} \varphi^{-1'}(w) > 0$. Hence $\varphi^{-1'}(w) \neq 0$ on $\partial\Delta$ and so $\partial[G] = \varphi^{-1}(\partial\Delta)$ is an analytic simple curve.

Let G be a domain as in Lemma 14.3. Let $x_1 = \inf_{x\in[G]\cap\mathbb{R}} x$ and $x_2 = \sup_{x\in[G]\cap\mathbb{R}} x$. Since $([G] \cap \mathbb{R})\setminus \operatorname{supp} \nu$ is open in $\mathbb{R}$, it is the union of open intervals I_j, $j = 1,2,\cdots$. Let I_1 be the interval such that $x_1 = \inf_{x\in I_1} x$ and I_2 be the interval such that $x_2 = \sup_{x\in I_2} x$. By using this notation, we have the following two lemmas:

Lemma 14.4. The domain $[G]$ is contained in $\{z \in \mathbb{C}|$

$x_1 < \mathrm{Re}\, z < x_2\}$ and $\partial[G]$ is tangent to the lines $\{z \in \mathbb{C} |$ $\mathrm{Re}\, z = x_1\}$ and $\{z \in \mathbb{C} | \mathrm{Re}\, z = x_2\}$ at $z = x_1$ and $z = x_2$, respectively.

Proof. Let φ be the one-to-one conformal mapping from [G] onto the unit disc Δ considered in the proof of Lemma 14.3. As in Lemma 14.3 take ζ_0 in I_1. Then the coefficients a_j of

$$\varphi^{-1}(w) = \zeta_0 + \frac{w}{\pi}\int \frac{\eta}{\eta - w} \frac{1}{\varphi^{-1'}(\eta^*)} d\mu \circ \psi^{-1}(\eta)$$

$$= \zeta_0 + \sum_{j=1}^{\infty} a_j w^j$$

are all positive. Hence $\mathrm{Re}\, \varphi^{-1}(w) = \zeta_0 + \Sigma a_j \mathrm{Re}\, w^j < \zeta_0 + \Sigma a_j = \varphi^{-1}(1) = x_2$ on Δ so that $[G] \subset \{z \in \mathbb{C} | \mathrm{Re}\, z < x_2\}$ and ∂G is tangent to $\{z \in \mathbb{C} | \mathrm{Re}\, z = x_2\}$ at $z = x_2$. The same holds for $\{z \in \mathbb{C} | \mathrm{Re}\, z = x_1\}$ and we obtain the lemma.

Lemma 14.5. There are no points z satisfying $\hat{\nu}(z) = \hat{\chi}_\Omega(z)$ on $I_1 \cup I_2$. There is at most one point z satisfying $\hat{\nu}(z) = \hat{\chi}_\Omega(z)$ on each I_j, $j \geq 3$.

Proof. A point $z \in [G]$ satisfies $\hat{\nu}(z) = \hat{\chi}_\Omega(z)$ if and only if $\frac{1}{\pi}(\hat{\chi}_{[G]}(z) + \pi\bar{z}) - \frac{1}{\pi}\hat{\mu}(z) = \bar{z}$. Hence this is equivalent to $f(w^*) = f(w)$ for $w = \varphi(z)$, where $f = \varphi^{-1}$ is the function defined before Proposition 10.19. We have already seen that $\hat{\nu} \neq \hat{\chi}_\Omega$ on $[G] \setminus \mathbb{R}$. Therefore we may assume $w \in \mathbb{R}$. Let w_1 and $w_2 (> w_1)$ be points on a connected component of $\mathbb{R} \setminus \psi(\mathrm{supp}\, \mu)$. Then $f(w_2) - f(w_1) > 0$, because $f'(w) > 0$ on $\mathbb{R} \setminus \psi(\mathrm{supp}\, \mu)$ by (14.1). Hence f is strictly increasing on each connected component of $\mathbb{R}\setminus\psi(\mathrm{supp}\ \mu)$. Since w^* decreases as $w > 0$ (or $w < 0$) increases, the lemma follows immediately.

Next we show that a finite positive measure ν with compact support can be approximated from below by measures ν_n satisfying $\inf_{z\in\mathrm{supp}\,\nu_n}\lambda(z;\nu_n,N) > 0$ if ν is singular with respect to m.

Lemma 14.6. Let ν be a finite positive measure with compact support. If ν is singular with respect to m, then there is a family $\{\nu(t)\}_{0\le t<M}$ of positive measures with compact supports such that $\nu(s)(E) < \nu(t)(E)$ for every pair of numbers s and t with $0 \le s < t$ and every Borel set E with $\nu(s)(E) > 0$, $\nu(t) \uparrow \nu$ as $t \uparrow M$ and $\inf_{z\in\mathrm{supp}\,\nu(t)}\lambda(z;\nu(t),N) > 0$ for every t with $0 \le t < M$.

Proof. Let $\lambda(z) = \lambda(z;\nu,4N)$ and $M = \max_{z\in\mathbb{C}}\lambda(z)$. For ε with $0 < \varepsilon \le M$, set $F_\varepsilon = \{z \in \mathbb{C} \mid \lambda(z) \ge \varepsilon\}$, $K_\varepsilon = \cup_{z\in F_\varepsilon}\overline{\Delta(\lambda(z);z)}$ and $\nu(M-\varepsilon) = \nu|K_\varepsilon$. Then $\lambda(z;\nu-\nu(M-\varepsilon),4N) < \varepsilon$ on $\mathbb{C}$. Since ν is singular, $\nu - \nu(M-\varepsilon) \downarrow 0$ as $\varepsilon \downarrow 0$. Since F_ε is compact, K_ε is also compact so that $\mathrm{supp}\,\nu(M-\varepsilon) \subset K_\varepsilon$. Let $p \in K_\varepsilon$. Then $p \in \overline{\Delta(\lambda(z);z)}$ for some $z \in F_\varepsilon$. Hence

$$\begin{aligned}(\nu|K_\varepsilon)\left(\overline{\Delta(2\lambda(z);p)}\right) &\ge \nu\left(\overline{\Delta(\lambda(z);z)}\right)\\ &= 4N\pi\lambda(z)^2\\ &= N\pi(2\lambda(z))^2\end{aligned}$$

so that $\lambda(p;\nu(M-\varepsilon),N) \ge 2\lambda(z) \ge 2\varepsilon$. Therefore

$$\inf_{\operatorname{supp}\nu(M-\varepsilon)} \lambda(z;\nu(M-\varepsilon),N) \geq \inf_{K_\varepsilon} \lambda(z;\nu(M-\varepsilon),N) \geq 2\varepsilon > 0$$

for every ε with $0 < \varepsilon \leq M$. Thus $\{\nu(t)\}$ satisfies $\inf_{\operatorname{supp}\nu(t)}\lambda(z;\nu(t),N) > 0$ for every t with $0 \leq t < M$.

To construct a family with required property, we construct $\{\nu(t)\}$ as above starting from $\nu/2$ instead of from ν. Then $\{(1+(t/M))\nu(t)\}$ is the required family.

Proposition 14.7. Let ν be a finite positive measure on $\mathbb{R}$ with compact support K. Then there is a unique open set $\tilde{W}$ in $Q(\nu,SL^1)$. The open set $\tilde{W}$ is symmetric with respect to $\mathbb{R}$. Every connected component of $\tilde{W}$ is simply connected and its boundary is analytic on $\mathbb{R}^c$. If $\inf_{z\in K}\lambda(z;\nu,144) > 0$, then $K \subset \tilde{W}$, the number of connected components of $\tilde{W}$ is finite and the boundary of each connected component of $\tilde{W}$ is an analytic simple curve.

Proof. Let $N = 100e$ and let $\{\nu(t)\}_{0\leq t<M}$ be the family stated as in Lemma 14.6. Since $\inf_{\operatorname{supp}\nu(t)}\lambda(z;\nu(t),100e) > 0$, by Lemma 14.1, there is the minimum open set $\tilde{W}(t) \in Q(\nu(t),SL^1)$ for every t with $0 \leq t < M$. Hence $\tilde{W} = \cup_{0\leq t<M} \tilde{W}(t) \in Q(\nu,SL^1)$ and $\tilde{W}$ is symmetric with respect to $\mathbb{R}$. Let F be an arbitrary compact connected subset of $\tilde{W}$. Then $F \subset G$ for some connected component G of some $\tilde{W}(t)$. By Lemma 14.3, $[G]$ is simply connected and, by an argument similar to Proposition 3.8, $[G] \subset \tilde{W}(t')$ for t' with $t < t' < M$. Hence every connected component of $\tilde{W}$ is simply connected and its boundary intersects with $\mathbb{R}$ at two points. By using an argument similar to Lemma 14.3, we see that the boundary is analytic on $\mathbb{R}^c$. Thus $[\tilde{W}] = \tilde{W}$ so that $\tilde{W}$ is unique in $Q(\nu,SL^1)$.

If $\inf_{z\in K}\lambda(z;\nu,144) > 0$, then we can construct an open set belonging to $Q(\nu,SL^1)$. By the uniqueness it is equal to $\tilde{W}$. Hence the proposition follows from Lemma 14.3.

For the quadrature domains for classes HL^1 and AL^1 we have

Proposition 14.8. Let ν be as in Proposition 14.7. Then $Q(\nu,HL^1) = Q(\nu,SL^1)$, namely, there is a unique open set $\tilde{W}$ in $Q(\nu,HL^1)$. If $\Omega \in Q(\nu,AL^1)$, then $[\Omega] = \tilde{W}$ and $\tilde{W}\backslash\Omega \subset \mathbb{R}$. If $\Omega \in Q(\nu,AL^1)$ contains supp ν, then $\tilde{W}\backslash\Omega$ is a finite set and there is at most one point of $\tilde{W}\backslash\Omega$ on each connected component I of $(\tilde{W} \cap \mathbb{R})\backslash\text{supp }\nu$. If $I \cap (\tilde{W}\backslash\Omega) \neq \phi$, then $\bar{I}\backslash I \subset \text{supp }\nu$.

Proof. At first we shall show that if $\Omega \in Q(\nu,AL^1)$ contains L, where L is the least closed interval containing supp ν and if $\tilde{W} \supset \text{supp }\nu$, then $\Omega = \tilde{W}$. We apply Proposition 9.4 replacing Ω and G by $\tilde{W}$ and Ω, respectively. By Lemma 14.2, $E(\tilde{W};\nu,AL^1)\backslash\Omega = \phi$ so that $\tilde{W}^e \cap \Omega = \phi$. Hence $\Omega \subset (\bar{\tilde{W}})^\circ = \tilde{W}$ so that $\Omega = \tilde{W}$.

Next we shall show that if $\Omega \in Q(\nu,AL^1)$, then $[\Omega] = \tilde{W}$ and $\tilde{W}\backslash\Omega \subset \mathbb{R}$. Set $d\mu = \chi_L dx$ and let $\{\tilde{W}(t)\}$ and $\{\Omega(t)\}$ be the minimum family in $Q(\{\chi_{\tilde{W}}m+t\mu\},SL^1)$ and $Q(\{\chi_\Omega m+t\mu\},SL^1)$, respectively. If $t > 0$, then $\text{supp}(\nu+t\mu) = L$. Hence, by the above argument, we have $\tilde{W}(t) = \Omega(t)$ for $t > 0$. Therefore $[\Omega] = [\tilde{W}] = \tilde{W}$. Since $\tilde{W}$ is symmetric with respect to $\mathbb{R}$, by Lemma 14.2, we obtain $\tilde{W}\backslash\Omega \subset \mathbb{R}$.

The assertion for the case that supp $\nu \subset \Omega$ follows from

Lemma 14.5.

Finally we show $Q(\nu,HL^1) = Q(\nu,SL^1)$. If $\Omega \in Q(\nu,HL^1)$, then $\Omega \in Q(\nu,AL^1)$. Hence $[\Omega] = \tilde{W}$. By using an argument similar to Theorem 7.5, we have $\tilde{W} \subset \Omega$. Therefore $\Omega = \tilde{W}$.

Let G be connected component of $\tilde{W}$. Let $x_1 = \inf_{x \in G \cap \mathbb{R}} x$ and $x_2 = \sup_{x \in G \cap \mathbb{R}} x$. Then $\{x_1, x_2\} = (\partial G) \cap \mathbb{R}$ and, by Lemma 14.4, $G \subset \{z \in \mathbb{C} \mid x_1 < \operatorname{Re} z < x_2\}$. Hence, if we define $\theta(x_1,G)$ and $\theta(x_2,G)$ by

$$\theta(x_1,G) = 2 \limsup_{\partial G \ni p \to x_1} \arg(p-x_1),$$

$$\theta(x_2,G) = 2\pi - 2 \liminf_{\partial G \ni p \to x_2} |\arg(p-x_2)|,$$

where $-\pi < \arg(p-x_i) < \pi$, then $\theta(x_i,G) \leq \pi$ for $i = 1,2$.

Concerning $\theta(x,G)$ we have

<u>Proposition 14.9</u>. Let ν be a finite positive measure on $[0,1]$ such that $d\nu \geq kx^\alpha dx$ on $[0,\delta]$, where $0 < \delta \leq 1$, k is a positive constant and α is a constant with $\alpha > -1$. Let $\tilde{W}$ be the maximum open set in $Q(\nu,AL^1)$. If $\alpha < 1$, or if $\alpha = 1$ and $k > \pi$, then $0 \in \tilde{W}$. If $\alpha = 1$, $k \leq \pi$ and $0 \notin \tilde{W}$, then $k \leq \theta(0,G)$ for a connected component G of $\tilde{W}$ with $0 \in \partial G$.

<u>Proof</u>. Assume $0 \notin \tilde{W}$ and set $\theta = \theta(0,G)$ for a connected component G of $\tilde{W}$ with $0 \in \partial G$. For a sufficiently small $\varepsilon > 0$, set $s_\varepsilon = \max\{\varepsilon/|z|-1, 0\}$. Then $s_\varepsilon \in SL^1(\tilde{W})$ so that $\int s_\varepsilon \, d\nu \leq \int_{\tilde{W}} s_\varepsilon dm$. Since

$$\int s_\varepsilon d\nu \geq \int_0^\varepsilon s_\varepsilon(kx^\alpha)dx = \varepsilon k\int_0^\varepsilon x^{\alpha-1}dx - k\int_0^\varepsilon x^\alpha dx,$$

we have $\alpha > 0$. If $\alpha > 0$, then the right-hand side of the above inequality is equal to $k\varepsilon^{\alpha+1}/(\alpha(\alpha+1))$. Since

$$\int_{\tilde{W}} s_\varepsilon dm \leq \frac{\theta}{2}\varepsilon^2 + o(\varepsilon^2),$$

by letting $\varepsilon \to 0$, we have $\alpha > 1$ or $\alpha = 1$ and $k \leq \theta$. The proposition follows from the fact that $\theta \leq \pi$.

Example 14.10. Let ω be the measure on $[-1,1]$ defined by $d\omega(x) = 2(1-\sqrt{|x|})dx$. Then $\tilde{W} = \{z = x+iy \mid -1 < x < 1,\ x^2/2 - 1/2 < y < -x^2/2 + 1/2\}$ is the unique domain in $Q(\omega, AL^1)$. Set $x = -1 + t$. Then $d\omega \geq tdt$ for $t \in [0,1]$. By the notation in Proposition 14.9, $\alpha = k = 1$. It is easy to show that $\theta(\pm 1, \tilde{W}) = \pi/2$. To show $\tilde{W} \in Q(\omega, AL^1)$, it is enough to see that

$$\int_{\tilde{W}} z^n dm = 2\int_{-1}^1 x^n(1-\sqrt{|x|})dx$$

for every integer $n \geq 0$. This follows from the following:

$$\int_{\tilde{W}} z^n dm = \int_{-1}^1 \left\{\int_{-f(x)}^{f(x)} z^n dy\right\} dx, \qquad \left(f(x) = -\frac{x^2}{2} + \frac{1}{2}\right)$$

$$\int_{-f(x)}^{f(x)} z^n dy = \frac{1}{i}\int_{-f(x)}^{f(x)} z^n d(iy) = \frac{1}{i}\frac{1}{n+1} z^{n+1} \Big|_{x-if(x)}^{x+if(x)},$$

$$\int_{-1}^1 (x+if(x))^{n+1}dx = \int_{-1}^1 \{(-\tfrac{i}{2})(x+i)^2\}^{n+1}dx$$

$$= \left(-\frac{i}{2}\right)^{n+1} \frac{1}{2n+3} z^{2n+3} \Big|_{-1+i}^{1+i}$$

$$= \frac{1}{2n+3} \{(1+i) - (-1)^{n+1} i\},$$

$$\int_{\tilde{W}} z^n dm = \frac{1}{i} \frac{1}{n+1} \cdot 2i \operatorname{Im} \frac{1}{2n+3} \{(1+i) - (-1)^{n+1} i\}$$

$$= 0 \quad (\text{for odd } n), \quad \frac{4}{(n+1)(2n+3)} \quad (\text{for even } n).$$

Let ν be a finite positive measure on $\mathbb{R}$ with compact support. In Proposition 14.7 we have proved that there is a unique open set $\tilde{W}$ in $Q(\nu, SL^1)$. By Proposition 14.8 it is the maximum open set in $Q(\nu, AL^1)$. Let us denote it by $Q(\nu)$. Next we shall show that if $Q(\nu) = Q(\mu)$, then $\nu = \mu$. More generally, we have

Proposition 14.11. Let μ and ν be complex measures on $\mathbb{C}$. If $K = \operatorname{supp} \mu \cup \operatorname{supp} \nu$ is a compact set such that K^c is connected and $K^\circ = \phi$, and if $Q(\mu, AL^1) \cap Q(\nu, AL^1) \neq \phi$, then $\mu = \nu$.

Proof. Let $\Omega \in Q(\mu, AL^1) \cap Q(\nu, AL^1)$. Then Ω is bounded and so

$$\int p d\mu = \int_\Omega p dm = \int p d\nu$$

for every polynomial p. Since the class of all polynomials is dense in the class of continuous functions on K, we have $\mu = \nu$.

Proposition 14.12. Let ν be a complex measure on $\mathbb{R}$ with compact support. If $\Omega \in Q(\nu, AL^1)$ is symmetric with respect to $\mathbb{R}$, then ν is real.

Proof. Since Ω is bounded, for every polynomial p with real coefficients, it follows that

$$\int p d\nu = \int_{\Omega} p dm = \int_{\Omega} \bar{p} dm.$$

Hence $\int p d(\mathrm{Im}\,\nu) = 0$ for every polynomial p with real coefficients and so $\mathrm{Im}\,\nu = 0$.

For the case when ν is of the form $\Sigma_{j=1}^{n} a_j \delta_{x_j}$, where δ_{x_j} denotes the Dirac measure at $x_j \in \mathbb{R}$, Proposition 14.12 was proved by Ullemar [23].

Now let us consider "quadratures". We say measures μ and ν satisfies the quadrature relation for class F if $Q(\mu,F) \cap Q(\nu,F) \neq \phi$ and denote it by $\mu Q(F)\nu$. In Chapter I, we have mainly concerned with the construction of a domain Ω in $Q(\mu,F)$ for a given μ. If we can find a measure ν such that $\Omega \in Q(\nu,F)$, then $\mu Q(F)\nu$. The better quadrature is the construction of measure ν which is easier to calculate the integral $\int f d\nu$ for $f \in F(\Omega)$.

We note here that if μ and ν are as in Proposition 14.11, then $\mu Q(AL^1)\nu$ implies $\mu = \nu$.

One of the construction method of measure ν satisfying $\Omega \in Q(\nu,SL^1)$ for a given Ω is obtained by using the Vitali covering theorem stated as at the beginning of this section. Another method has been given by Stenger [20] for a simply connected domain Ω.

BIBLIOGRAPHY

[1] D. Aharonov and H. S. Shapiro, Domains on which analytic functions satisfy quadrature identities, J. Anal. Math. 30 (1976), 39-73.

[2] D. Aharonov and H. S. Shapiro, A minimal-area problem in conformal mapping, Royal Institute of Technology research bulletin, 1973. Third printing 1978.

[3] Y. Avci, Quadrature identities and the Schwarz function, Dissertation, Stanford university, 1977.

[4] S. Bergman, The kernel function and conformal mapping, Math. Surveys V, Amer. Math. Soc., sec. ed., 1970.

[5] L. Bers, An approximation theorem, J. Anal. Math. 14 (1965), 1-4.

[6] P. J. Davis, Simple quadratures in the complex plane, Pacific J. Math. 15 (1965), 813-824.

[7] P. J. Davis, Additional simple quadratures in the complex plane, Aequationes Math. 3 (1969), 149-155.

[8] P. J. Davis, Double integrals expressed as single integrals or interpolatory functionals, J. Approximation Theory 5 (1972), 276-307.

[9] P. J. Davis, The Schwarz function and its applications, Carus Math. Monographs, No. 17, Math. Assoc. Amer., 1974.

[10] J. Garnett, Analytic capacity and measure, Lecture Notes i Math. No. 297, Springer, Berlin, 1972.

[11] B. Gustafsson, Quadrature identities and the Schottky

double, Royal Institute of Technology research bulletin, 1977.

[12] L. I. Hedberg, Approximation in the mean by solutions of elliptic equations, Duke Math. J. 40 (1973), 9-16.

[13] Ü. Kuran, On the mean-value property of harmonic functions, Bull. London Math. Soc. 4 (1972), 311-312.

[14] H. Lamb, Hydrodynamics, Cambridge Univ. Press, London, 1932.

[15] S. Richardson, Hell Shaw flows with a free boundary produced by the injection of fluid into a narrow channel. J. Fluid Mech. 56 (1972), 609-618.

[16] M. Sakai, On basic domains of extremal functions, Kōdai Math. Sem. Rep. 24 (1972), 251-258.

[17] M. Sakai, Analytic functions with finite Dirichlet integrals on Riemann surfaces, Acta Math. 142 (1979), 199-220.

[18] M. Sakai, The sub-mean-value property of subharmonic functions and its application to the estimation of the Gaussian curvature of the span metric, Hiroshima Math. J. 9 (1979), 555-593.

[19] M. Schiffer and D. C. Spencer, Functionals of finite Riemann surfaces, Princeton Univ. Press, Princeton, 1954.

[20] F. Stenger, The reduction of two dimensional integrals into a finite number of one dimensional integrals, Aequationes Math. 6 (1971), 278-287.

[21] G. Szegö, Orthogonal polynomials, Amer. Math. Soc. Colloq.

Publ. Vol. XXIII, Providence, 1939.

[22] M. Tsuji, Potential theory in modern function theory, Chelsea, New York, sec. ed., 1975.

[23] C. Ullemar, Symmetric plane domains satisfying two-point quadrature identities for analytic functions, Royal Institute of Technology research bulletin, 1977.

LIST OF SYMBOLS

$\mathbb{C}$ is the complex plane (or the set of complex numbers).

$\mathbb{R}$ is the set of real numbers.

ϕ is the empty set.

A^c is the complement of A in $\mathbb{C}$.

A^e is the exterior of A.

A° is the interior of A.

$\bar{A}$ is the closure of A if A is a subset of $\mathbb{C}$.

∂A is the boundary of A.

$A \setminus B$ is the difference between A and B, namely, $A \setminus B = A \cap B^c$.

$A \Delta B$ is the symmetric difference of A and B, namely, $A \Delta B = (A \setminus B) \cup (B \setminus A)$.

$[\Omega]$ denotes the areal maximal domain of Ω(see p. 4).

$[\Omega]^{cap}$ denotes the maximal domain of Ω with respect to capacity (see p. 32).

$\Omega \oplus d$ is defined on p. 71.

$\Omega \ominus d$ is defined on p. 71.

$\Delta(r;c)$ denotes the open disc with radius r and center at c. In §10 we abbreviate $\Delta(r;0)$ to $\Delta(r)$.

(r;c) denotes the closed square with horizontal and vertical sides of length 2r and center at c.

A,B) is the distance between A and B.

is the characteristic function of A.

is the restriction of a mapping (or a measure) ν onto A.

$\ell(\gamma)$ denotes the length of an arc γ.

m_1 denotes the one-dimensional Lebesgue measure.

m denotes the two-dimensional Lebesgue measure.

$\bar{z}$ is the complex conjugate of z if z is a complex number.

$f \circ g$ is the composite of mappings f and g.

$\nu \circ \varphi^{-1}$ is the measure on Y defined by $(\nu \circ \varphi^{-1})(E) = \nu(\varphi^{-1}(E))$ for every Borel set E in Y, where φ is a measurable transformation from X to Y and ν is a measure on X.

$\hat{\nu}$ denotes the Cauchy transform of a measure ν (see p. 4).

U^{ν} is defined on p. 62.

$\|\nu\|$ is the total variation of a measure ν.

$d\nu/dm$ is the Radon-Nikodym derivative of ν.

$AL^p(\Omega)$, where Ω is an open set in $\mathbb{C}$, denotes the class of analytic L^p-functions on Ω.

$HL^p(\Omega)$ denotes the class of harmonic L^p-functions on Ω.

$SL^p(\Omega)$ denotes the class of subharmonic L^p-functions on Ω.

$Q(\nu,F)$ denotes the class of all quadrature domains of ν with finite area for class F (see p. 2).

$Q_\infty(\nu,F)$ denotes the class of all quadrature domains of ν with infinite area for class F (see p. 2).

$n(p)$ is defined on p. 9.

angle V_j is defined on p. 9.

$\lambda(z;\mu,N)$ is defined on p. 10.

$\lambda_S(z;\mu,N)$ is defined on p. 34.

$\beta(\mu,W)$ is defined on p. 16.

$f(z;\mu,\Delta(r;c))$ is defined on p. 18.

$E(\Omega;\nu,AL^1)$ is defined on p. 58.

$E(\Omega;\nu,HL^1)$ is defined on p. 58.

$Q(\{\nu(t)\},F)$ is defined on p. 71.

$\mathrm{disc}\{\Omega(t)\}$ is defined on p. 72.

$\mathrm{stag}\{\Omega(t)\}$ is defined on p. 73.

$\mathrm{stat}\{\Omega(t)\}$ is defined on p. 73.

INDEX

ɔl. 787: Potential Theory, Copenhagen 1979. Proceedings, 1979. dited by C. Berg, G. Forst and B. Fuglede. VIII, 319 pages. 1980.

ol. 788: Topology Symposium, Siegen 1979. Proceedings, 1979. dited by U. Koschorke and W. D. Neumann. VIII, 495 pages. 1980.

Vol. 789: J. E. Humphreys, Arithmetic Groups. VII, 158 pages. 1980.

Vol. 790: W. Dicks, Groups, Trees and Projective Modules. IX, 127 pages. 1980.

Vol. 791: K. W. Bauer and S. Ruscheweyh, Differential Operators for Partial Differential Equations and Function Theoretic Applications. V, 258 pages. 1980.

Vol. 792: Geometry and Differential Geometry. Proceedings, 1979. Edited by R. Artzy and I. Vaisman. VI, 443 pages. 1980.

Vol. 793: J. Renault, A Groupoid Approach to C*-Algebras. III, 160 pages. 1980.

Vol. 794: Measure Theory, Oberwolfach 1979. Proceedings 1979. Edited by D. Kölzow. XV, 573 pages. 1980.

Vol. 795: Séminaire d'Algèbre Paul Dubreil et Marie-Paule Malliavin. Proceedings 1979. Edited by M. P. Malliavin. V, 433 pages. 1980.

Vol. 796: C. Constantinescu, Duality in Measure Theory. IV, 197 pages. 1980.

Vol. 797: S. Mäki, The Determination of Units in Real Cyclic Sextic Fields. III, 198 pages. 1980.

Vol. 798: Analytic Functions, Kozubnik 1979. Proceedings. Edited by J. Ławrynowicz. X, 476 pages. 1980.

Vol. 799: Functional Differential Equations and Bifurcation. Proceedings 1979. Edited by A. F. Izé. XXII, 409 pages. 1980.

Vol. 800: M.-F. Vignéras, Arithmétique des Algèbres de Quaternions. VII, 169 pages. 1980.

Vol. 801: K. Floret, Weakly Compact Sets. VII, 123 pages. 1980.

Vol. 802: J. Bair, R. Fourneau, Etude Géometrique des Espaces Vectoriels II. VII, 283 pages. 1980.

Vol. 803: F.-Y. Maeda, Dirichlet Integrals on Harmonic Spaces. X, 180 pages. 1980.

Vol. 804: M. Matsuda, First Order Algebraic Differential Equations. VII, 111 pages. 1980.

Vol. 805: O. Kowalski, Generalized Symmetric Spaces. XII, 187 pages. 1980.

Vol. 806: Burnside Groups. Proceedings, 1977. Edited by J. L. Mennicke. V, 274 pages. 1980.

ɔl. 807: Fonctions de Plusieurs Variables Complexes IV. Proceedings, 1979. Edited by F. Norguet. IX, 198 pages. 1980.

808: G. Maury et J. Raynaud, Ordres Maximaux au Sens de sano. VIII, 192 pages. 1980.

809: I. Gumowski and Ch. Mira, Recurences and Discrete mic Systems. VI, 272 pages. 1980.

10: Geometrical Approaches to Differential Equations. Progs 1979. Edited by R. Martini. VII, 339 pages. 1980.

1: D. Normann, Recursion on the Countable Functionals. pages. 1980.

Y. Namikawa, Toroidal Compactification of Siegel Spaces. pages. 1980.

A. Campillo, Algebroid Curves in Positive Characteristic. es. 1980.

éminaire de Théorie du Potentiel, Paris, No. 5. Proceed- l by F. Hirsch et G. Mokobodzki. IV, 239 pages. 1980.

J. Slodowy, Simple Singularities and Simple Algebraic 75 pages. 1980.

Stoica, Local Operators and Markov Processes. VIII, 80.

Vol. 817: L. Gerritzen, M. van der Put, Schottky Groups and Mumford Curves. VIII, 317 pages. 1980.

Vol. 818: S. Montgomery, Fixed Rings of Finite Automorphism Groups of Associative Rings. VII, 126 pages. 1980.

Vol. 819: Global Theory of Dynamical Systems. Proceedings, 1979. Edited by Z. Nitecki and C. Robinson. IX, 499 pages. 1980.

Vol. 820: W. Abikoff, The Real Analytic Theory of Teichmüller Space. VII, 144 pages. 1980.

Vol. 821: Statistique non Paramétrique Asymptotique. Proceedings, 1979. Edited by J.-P. Raoult. VII, 175 pages. 1980.

Vol. 822: Séminaire Pierre Lelong–Henri Skoda, (Analyse) Années 1978/79. Proceedings. Edited by P. Lelong et H. Skoda. VIII, 356 pages, 1980.

Vol. 823: J. Král, Integral Operators in Potential Theory. III, 171 pages. 1980.

Vol. 824: D. Frank Hsu, Cyclic Neofields and Combinatorial Designs. VI, 230 pages. 1980.

Vol. 825: Ring Theory, Antwerp 1980. Proceedings. Edited by F. van Oystaeyen. VII, 209 pages. 1980.

Vol. 826: Ph. G. Ciarlet et P. Rabier, Les Equations de von Kármán. VI, 181 pages. 1980.

Vol. 827: Ordinary and Partial Differential Equations. Proceedings, 1978. Edited by W. N. Everitt. XVI, 271 pages. 1980.

Vol. 828: Probability Theory on Vector Spaces II. Proceedings, 1979. Edited by A. Weron. XIII, 324 pages. 1980.

Vol. 829: Combinatorial Mathematics VII. Proceedings, 1979. Edited by R. W. Robinson et al.. X, 256 pages. 1980.

Vol. 830: J. A. Green, Polynomial Representations of GL_n. VI, 118 pages. 1980.

Vol. 831: Representation Theory I. Proceedings, 1979. Edited by V. Dlab and P. Gabriel. XIV, 373 pages. 1980.

Vol. 832: Representation Theory II. Proceedings, 1979. Edited by V. Dlab and P. Gabriel. XIV, 673 pages. 1980.

Vol. 833: Th. Jeulin, Semi-Martingales et Grossissement d'une Filtration. IX, 142 Seiten. 1980.

Vol. 834: Model Theory of Algebra and Arithmetic. Proceedings, 1979. Edited by L. Pacholski, J. Wierzejewski, and A. J. Wilkie. VI, 410 pages. 1980.

Vol. 835: H. Zieschang, E. Vogt and H.-D. Coldewey, Surfaces and Planar Discontinuous Groups. X, 334 pages. 1980.

Vol. 836: Differential Geometrical Methods in Mathematical Physics. Proceedings, 1979. Edited by P. L. García, A. Pérez-Rendón, and J. M. Souriau. XII, 538 pages. 1980.

Vol. 837: J. Meixner, F. W. Schäfke and G. Wolf, Mathieu Functions and Spheroidal Functions and their Mathematical Foundations Further Studies. VII, 126 pages. 1980.

Vol. 838: Global Differential Geometry and Global Analysis. Proceedings 1979. Edited by D. Ferus et al. XI, 299 pages. 1981.

Vol. 839: Cabal Seminar 77 – 79. Proceedings. Edited by A. S. Kechris, D. A. Martin and Y. N. Moschovakis. V, 274 pages. 1981.

Vol. 840: D. Henry, Geometric Theory of Semilinear Parabolic Equations. IV, 348 pages. 1981.

Vol. 841: A. Haraux, Nonlinear Evolution Equations- Global Behaviour of Solutions. XII, 313 pages. 1981.

Vol. 842: Séminaire Bourbaki vol. 1979/80. Exposés 543–560. IV, 317 pages. 1981.

Vol. 843: Functional Analysis, Holomorphy, and Approximation Theory. Proceedings. Edited by S. Machado. VI, 636 pages. 1981.

Vol. 844: Groupe de Brauer. Proceedings. Edited by M. Kervaire and M. Ojanguren. VII, 274 pages. 1981.

Vol. 845: A. Tannenbaum, Invariance and System Theory: Algebraic and Geometric Aspects. X, 161 pages. 1981.

Vol. 846: Ordinary and Partial Differential Equations, Proceedings. Edited by W. N. Everitt and B. D. Sleeman. XIV, 384 pages. 1981.

Vol. 847: U. Koschorke, Vector Fields and Other Vector Bundle Morphisms – A Singularity Approach. IV, 304 pages. 1981.

Vol. 848: Algebra, Carbondale 1980. Proceedings. Ed. by R. K. Amayo. VI, 298 pages. 1981.

Vol. 849: P. Major, Multiple Wiener-Itô Integrals. VII, 127 pages. 1981.

Vol. 850: Séminaire de Probabilités XV. 1979/80. Avec table générale des exposés de 1966/67 à 1978/79. Edited by J. Azéma and M. Yor. IV, 704 pages. 1981.

Vol. 851: Stochastic Integrals. Proceedings, 1980. Edited by D. Williams. IX, 540 pages. 1981.

Vol. 852: L. Schwartz, Geometry and Probability in Banach Spaces. X, 101 pages. 1981.

Vol. 853: N. Boboc, G. Bucur, A. Cornea, Order and Convexity in Potential Theory: H-Cones. IV, 286 pages. 1981.

Vol. 854: Algebraic K-Theory. Evanston 1980. Proceedings. Edited by E. M. Friedlander and M. R. Stein. V, 517 pages. 1981.

Vol. 855: Semigroups. Proceedings 1978. Edited by H. Jürgensen, M. Petrich and H. J. Weinert. V, 221 pages. 1981.

Vol. 856: R. Lascar, Propagation des Singularités des Solutions d'Equations Pseudo-Différentielles à Caractéristiques de Multiplicités Variables. VIII, 237 pages. 1981.

Vol. 857: M. Miyanishi. Non-complete Algebraic Surfaces. XVIII, 244 pages. 1981.

Vol. 858: E. A. Coddington, H. S. V. de Snoo: Regular Boundary Value Problems Associated with Pairs of Ordinary Differential Expressions. V, 225 pages. 1981.

Vol. 859: Logic Year 1979–80. Proceedings. Edited by M. Lerman, J. Schmerl and R. Soare. VIII, 326 pages. 1981.

Vol. 860: Probability in Banach Spaces III. Proceedings, 1980. Edited by A. Beck. VI, 329 pages. 1981.

Vol. 861: Analytical Methods in Probability Theory. Proceedings 1980. Edited by D. Dugué, E. Lukacs, V. K. Rohatgi. X, 183 pages. 1981.

Vol. 862: Algebraic Geometry. Proceedings 1980. Edited by A. Libgober and P. Wagreich. V, 281 pages. 1981.

Vol. 863: Processus Aléatoires à Deux Indices. Proceedings, 1980. Edited by H. Korezlioglu, G. Mazziotto and J. Szpirglas. V, 274 pages. 1981.

Vol. 864: Complex Analysis and Spectral Theory. Proceedings, 1979/80. Edited by V. P. Havin and N. K. Nikol'skii. VI, 480 pages. 1981.

Vol. 865: R. W. Bruggeman, Fourier Coefficients of Automorphic Forms. III, 201 pages. 1981.

Vol. 866: J.-M. Bismut, Mécanique Aléatoire. XVI, 563 pages. 1981.

Vol. 867: Séminaire d'Algèbre Paul Dubreil et Marie-Paule Malliavin. Proceedings, 1980. Edited by M.-P. Malliavin. V, 476 pages. 1981.

Vol. 868: Surfaces Algébriques. Proceedings 1976–78. Edited by J. Giraud, L. Illusie et M. Raynaud. V, 314 pages. 1981.

Vol. 869: A. V. Zelevinsky, Representations of Finite Classical Groups. IV, 184 pages. 1981.

Vol. 870: Shape Theory and Geometric Topology. Proceedings, 1981. Edited by S. Mardešić and J. Segal. V, 265 pages. 1981.

Vol. 871: Continuous Lattices. Proceedings, 1979. Edited by B. Banaschewski and R.-E. Hoffmann. X, 413 pages. 1981.

Vol. 872: Set Theory and Model Theory. Proceedings, 1979. Edited by R. B. Jensen and A. Prestel. V, 174 pages. 1981.

Vol. 873: Constructive Mathematics, Proceedings, 1980. Edited F. Richman. VII, 347 pages. 1981.

Vol. 874: Abelian Group Theory. Proceedings, 1981. Edited R. Göbel and E. Walker. XXI, 447 pages. 1981.

Vol. 875: H. Zieschang, Finite Groups of Mapping Classes c Surfaces. VIII, 340 pages. 1981.

Vol. 876: J. P. Bickel, N. El Karoui and M. Yor. Ecole d'Eté de Proba bilités de Saint-Flour IX – 1979. Edited by P. L. Hennequin. X 280 pages. 1981.

Vol. 877: J. Erven, B.-J. Falkowski, Low Order Cohomology anc Applications. VI, 126 pages. 1981.

Vol. 878: Numerical Solution of Nonlinear Equations. Proceedings 1980. Edited by E. L. Allgower, K. Glashoff, and H.-O. Peitgen XIV, 440 pages. 1981.

Vol. 879: V. V. Sazonov, Normal Approximation – Some Recen Advances. VII, 105 pages. 1981.

Vol. 880: Non Commutative Harmonic Analysis and Lie Groups Proceedings, 1980. Edited by J. Carmona and M. Vergne. IV, 55 pages. 1981.

Vol. 881: R. Lutz, M. Goze, Nonstandard Analysis. XIV, 261 pages 1981.

Vol. 882: Integral Representations and Applications. Proceedings 1980. Edited by K. Roggenkamp. XII, 479 pages. 1981.

Vol. 883: Cylindric Set Algebras. By L. Henkin, J. D. Monk, A. Tarski H. Andréka, and I. Németi. VII, 323 pages. 1981.

Vol. 884: Combinatorial Mathematics VIII. Proceedings, 1980. Edite by K. L. McAvaney. XIII, 359 pages. 1981.

Vol. 885: Combinatorics and Graph Theory. Edited by S. B. Rao Proceedings, 1980. VII, 500 pages. 1981.

Vol. 886: Fixed Point Theory. Proceedings, 1980. Edited by E. Fadel and G. Fournier. XII, 511 pages. 1981.

Vol. 887: F. van Oystaeyen, A. Verschoren, Non-commutative Alge braic Geometry, VI, 404 pages. 1981.

Vol. 888: Padé Approximation and its Applications. Proceedings, 1980. Edited by M. G. de Bruin and H. van Rossum. VI, 383 pages. 1981.

Vol. 889: J. Bourgain, New Classes of $\mathcal{L}^p$-Spaces. V, 143 pages. 1981

Vol. 890: Model Theory and Arithmetic. Proceedings, 1979/8 Edited by C. Berline, K. McAloon, and J.-P. Ressayre. VI, 30 pages. 1981.

Vol. 891: Logic Symposia, Hakone, 1979, 1980. Proceedings, 197 1980. Edited by G. H. Müller, G. Takeuti, and T. Tugué. XI, 394 pag 1981.

Vol. 892: H. Cajar, Billingsley Dimension in Probability Spa III, 106 pages. 1981.

Vol. 893: Geometries and Groups. Proceedings. Edited by M. A and D. Jungnickel. X, 250 pages. 1981.

Vol. 894: Geometry Symposium. Utrecht 1980, Proceedings. by E. Looijenga, D. Siersma, and F. Takens. V, 153 pages

Vol. 895: J.A. Hillman, Alexander Ideals of Links. V, 178 page

Vol. 896: B. Angéniol, Familles de Cycles Algébriques – Sc Chow. VI, 140 pages. 1981.

Vol. 897: W. Buchholz, S. Feferman, W. Pohlers, W. Sie Inductive Definitions and Subsystems of Analysis: Re Theoretical Studies. V, 383 pages. 1981.

Vol. 898: Dynamical Systems and Turbulence, War Proceedings. Edited by D. Rand and L.-S. Young. VI, 390

Vol. 899: Analytic Number Theory. Proceedings, 198 M.I. Knopp. X, 478 pages. 1981.